GOD AND THE MIND MACHINE

GOD AND THE MIND MACHINE

Computers, Artificial Intelligence
and the Human Soul

John Puddefoot

FOREWORD BY
JOHN POLKINGHORNE

'The computer is a powerful new metaphor for helping us to understand many aspects of the world, but ... it enslaves the mind that has no other metaphors and few other resources to call on.'

Joseph Weizenbaum[1]

First published 1996
SPCK
Holy Trinity Church
Marylebone Road
London
NW1 4DU

Copyright © John C. Puddefoot 1996

All rights reserved. No part of this book may be reproduced or
transmitted in any form or by any means, electronic or mechanical,
including photocopying, recording, or by any information storage
and retrieval system, without permission in writing
from the publisher.

British Library Cataloguing in Publication Data
A catalogue record for this book is available from the British Library.

ISBN 0–281–04973–4

Typeset by Wilmaset Ltd, Birkenhead, Wirral.
Printed in Great Britain by
Biddles Ltd, Guildford and Kings Lynn.

For Hilary

Contents

Foreword by John Polkinghorne	viii
Preface	ix
Acknowledgements	xii
Introduction	1
1 Dream or Fantasy?	7
2 Minds, Brains and Consciousness	39
3 Problems with Words	57
4 Computers, Theology and Science	76
5 God and the Mind Machine	95
6 The Universe, Everything and Life	124
Epilogue	133
Notes	137
Further Reading	139
Index	141

Foreword

What, if any, are the limits of Artificial Intelligence? Could a computer have a mind? Are we computers made of meat? (If so, that answers the preceding question!) What could be the meaning of the soul? Is it credible we have a destiny beyond our death? Will there be androids in heaven? Where does God fit into all this?

These are among the questions John Puddefoot addresses in this wide-ranging book. His style is clear, forthright, provocative and quizzical. His search is for an understanding of what he calls our 'inside-out' perspective (the subjectively experienced 'I-story' we each can tell) and how it relates to the 'outside-in' perspective (the objective 'it-story' of an external observer). These are very important issues and the author has many shrewd comments to make on them.

In such a complex and perplexing area of discourse, no two people are likely to agree on all the questions to ask, let alone on all the answers to be given. I do not agree with absolutely all of John Puddefoot's opinions, but I believe he presents them in a clear way which merits serious thought. I have greatly enjoyed reading the book and being his companion in a stimulating intellectual excursion. I am sure that many others will profit from a similar experience.

John Polkinghorne
Queens' College, Cambridge

Preface

The baby whose shining eyes adorn the cover of this book may have reached roughly my current age when I die. If she should then find herself sitting in front of a computer, what differences will there be between that machine and the one I am using to type these words?

Since we can already speak to computers and have them turn our spoken words into text (for example, using IBM's VoiceType^{TM} program), it is most unlikely that her computer will have a keyboard like mine. But that is merely a cosmetic change. Will the computer provide far more advanced functions? Will it act as a personal adviser about finance, health, education, career moves, as well as supplying automatic records of all financial transactions, piping music into chosen rooms in the house, and even babysitting, its sensors protecting children from fire, burglars and other dangers?

Such a machine will certainly play a better game of chess than any human being on earth. It will have access, through the descendants of the Internet, to all the knowledge bases in the world. The children of those days will probably be educated largely at home, spending time with others only for recreation, socialization and sport. Each home will have computers capable of tailoring the speed and difficulty of each educational topic to the needs and abilities of the children in its charge. There will be few examinations, if any. A child's test scores will be stored and accumulated to afford a complete academic and psychological profile that will be far more detailed than any examination. Continuous assessment will have come of age.

Do you feel a chill running down your spine as you read those words? Do some of the Hollywood B movies of the fifties and sixties echo in your memory, where computers take over the world, and human beings become subservient to machines? Or do you just dread the impersonal nature of the computer, and resent in horror the idea that a machine might be entrusted with the care and education of your children or grandchildren? Do you shudder at the thought of the control that could be exercised over young minds by those who control the sources and the packaging of information? Does the nightmare in which every child produces essays for school printed from identical copies of the latest CD-ROM encyclopaedia seem to have come true?

These are very real fears, and if advances in computer technology over recent years are anything to go by, these scenarios may be upon us more quickly than we think possible today. Computers are set to become the channels through which all human control and communication will take place; will they also control the way we think of ourselves, and the way in which and the extent to which we think of God?

Any book in the theology-science realm runs the risk of oversimplifying the issues which concern its readers. I have taken a stance in what follows in which I take seriously some of the challenges computer science poses to theology. I do not dismiss outright the possibility that computers may one day develop a form of consciousness, or that they may one day exceed us in intellectual power. I do not deny at the outset that the language of programs and machines, software and hardware, has influenced the way we think. Some no doubt both deny such claims and dismiss such language on the grounds that they do not concede, never have conceded, and never will concede that a computer is remotely like or capable of being remotely like a human being. They may appeal to the human spirit, mind or soul or to certain kinds of religious perspective to bolster their rejection of the entire problem.

I hope that such readers will bear with me, and allow me to explore the territory a little before they reject the problems out of hand. Christians have often been ready to say that something or other proposed by a scientific position is impossible on grounds of faith, Scripture or religious concern. In my view, a truly fundamental and unshakeable faith in God neither requires nor allows such a ready rejection. Just as Jesus entered into sin and a world of corruption and error in order to speak to and live with the lost, so religious people and theologians must be ready to enter into and explore territory that is deeply incompatible with their fundamental views and which they suspect involves practices and pursues objectives with which they are out of sympathy.

The world of computers is painted too darkly if represented in such terms, but there is little doubt that for many – especially those for whom computers are nothing more than a new-fangled gadget that repeatedly sends them incorrect electricity bills – every new development seems to lead us into darker and darker recesses of the human mind. The most recent development, the Internet, is perceived by some as a device for spreading pornography and gossip around the world. Widely-reported instances of universities and schools having to deny their students access to certain Internet sites – like the back-alley images that seem to preoccupy those who publish magazines on the same topics – serve only to underline and reinforce this suspicion.

So, readers who doubt that computers are a threat to theology may be disinclined to read much further, even if they have read this far. I hope that I shall be able to show not just that their basic instinct is right – that there is very little indeed to fear – but also that by allowing the newly emerging discipline of computer science to instruct us we can actually acquire significant new insights into what it is to be human (just by seeing and clarifying what it is a computer would need to achieve in order to be counted as such). In this way I hope to anticipate a good many of the arguments about computers that I think will occupy us in the near future, to forestall some of those who would dismiss them out of hand, and to set out some markers by which to tackle them.

Acknowledgements

As the following pages should make clear, a book owes more to the world which produced its author than to the author himself. Those who patiently and thanklessly taught him in school and university, who endured his endless theological and philosophical babblings when there must have been better ways to spend their time, and who lived with his tantrums when that extra-special idea just would not 'come', probably deserve more of the credit for this work than anyone else.

In particular I would like to express my unqualified gratitude to Tom Torrance for his support and encouragement over almost twenty years, to John Polkinghorne for his foreword, and the director, staff and trustees of the Center of Theological Inquiry, Princeton, for electing me to membership, and for two short periods in residence there during which the ideas which now find incomplete expression here began to germinate.

To the staff of SPCK, and especially to Rachel Boulding, who saw in an early draft of this book the potential for something which now bears little resemblance to what she then read; I owe my thanks for their encouragement, professionalism and forbearance. My thanks also go to Selma Thomas for her help in compiling the bibliographical references.

To my wife Hilary and Rachel, Helen and Hannah, my daughters, I owe apologies for my absence at the computer – still, when all is said and done, and despite all that is yet to be said, a far less agreeable companion than they – and thanks for their support, interest and much more besides.

John Puddefoot

Introduction

Theology and science

For much of its history, the debate about the relationship between theology and science has been characterized by aggressive science and defensive religion and theology. That continues to the present day, when theologians seem always to be at pains to point out the compatibility of theological and scientific views of the universe as complements to one another.

This is not the place to rehearse these arguments – the works in the bibliography by Tom Torrance, Ian Barbour, Arthur Peacocke, John Polkinghorne and others may be consulted for differing accounts of this rich discussion – but I would like to make my own position clear to avoid misunderstanding.

For me, the principal problem with discussions between theology and science is the persistent defensiveness of theologians in the face of a largely toothless scientific assault. There are, of course, countless scientists who reject all religion as superstitious nonsense and who employ their science to bolster such rejection, but a distinction must be drawn between what science has revealed of the world and how scientists choose (and I mean 'choose') to interpret what it says. An interpretation of science stands in relation to science as an interpretation of a picture stands to a picture; we should not allow the sometimes fluent atheism of some scientists to deny theology access to science. Atheistic scientists do not own science, and science is not inherently or necessarily atheistic (whatever those same fluent advocates may say). Sometimes, in fact, their florid and persuasive prose is the best clue to the thinness of the association between what they are saying and what science actually says. Such writers tend to talk about 'what science says' when they should speak of 'what they understand science to say', but they are seldom honest or open enough to admit the intrusion of their own preconceptions. And they prefer to paint with broad brush strokes where finer and more discerning lines are called for.

There seems to me to be no reason to be defensive about theology in the face of science, or to rest content merely with a shallow coexistence between the two. Neither will I concede, in the name of greater harmony between theology and science, a naturalization of God that reduces him to a feature of the universe. It is in my view essential that the God of whom we speak and in whom we believe would continue to exist even if the universe did not exist. I also welcome, almost without exception, all scientific and technological advance as a positive contribution both to our understanding of God's universe and to human progress (and I say this despite the fact that for many the sheer notion of progress is anathema).

I am absolutely sure that the religious stance that is rationally possible in a scientific world is superior to any such stance in any other era of human history, and that we are now able (if we care to ask and answer the questions) to understand divine creation, redemption and purpose far better than has ever been possible before. Not only many scientists, but many theologians and religious people will certainly disagree with me. Many will wish to hold on to the view that science can explain everything, or to the possibilities left open by superstition and mystery-mongering in certain kinds of religious belief. I am happy, even eager, to dispense with both.

But it would be unreasonable to claim that scientific advance had involved no loss at all, or that our religious sensibilities are as a matter of fact more finely tuned now than ever they were in the past. Much has been lost: our sense of the personal; our sense of genuine (as opposed to spurious) mystery; our sense of wonder and awe. Of course, many scientists profess the profoundest awe at the way the universe is made – an awe which I share and which computer science extends – but it is probably true that for many people that sense has been diminished by scientific advance. The task is therefore not to reject the past, but to *repossess* those aspects of it which we are the poorer for having lost. (In this regard it strikes me as useful to think of the way balanced adults have repossessed the innocence of the child and the rebelliousness of the adolescent, rather than rejecting both as things of the past for which they have no further use.)[1]

The structure of the book

This book is about the creatures God has made 'in his own image' – human beings – and the creatures some human beings aspire one day to make in theirs – artificially intelligent creatures – androids. It is about *machines, life, intelligence* and the human soul.

It is difficult to discuss these matters without first being clear about the ways we are going to use the words. Is artificial life alive? Is artificial intelligence intelligent? Is the brain a machine? Vagueness in our use of words may account for much of the confusion and uncertainty we experience over the whole field. Some scientists, for example, allow themselves the luxury of wanting to say that a motor car is alive. This seems to stretch the everyday meaning of 'alive' to the point where it becomes useless.

First we need to set the scene. What has been achieved? Is there a real question to answer? Ought religious people to be at all concerned about these changes? What may the future hold? How real are our science fantasies likely to become? Will we one day share our world with androids as intelligent and self-resourcing and useful as the amiable and benevolent Lieutenant Commander Data in *Star Trek*? Will we one day be overwhelmed by Cybermen, the rogue products of our own ambition? These matters and more will occupy Chapter 1: 'Dream or Fantasy?'

But if 'machine', 'life' and 'intelligence' are vague terms, how much more vague are *mind, soul* and *consciousness*? The existing literature is vast, complex and full of disagreements. Even to do justice to the disagreements would take many books, and would probably obscure the main point. I shall therefore set up a summary way of discussing these matters which I think centrally important to the clarification of the issues, but I cannot give as much attention as perhaps I should to the arguments against the position I shall adopt. This summary occupies the second chapter of the book. It is central to one of the principal arguments of the book – that to become a creature such as ourselves an android will need a mind, an inside-looking-out quality such as we each enjoy. We need to be clear about what this entails and what difficulties it creates. All this is in Chapter 2: 'Minds, Brains and Consciousness'.

We shall also need working definitions for machine, life and intelligence. While most of us would agree on certain things that must be satisfied before something can be described as a machine, as alive, or as intelligent, there are also additional properties which some think essential and others do not. So we find ourselves saying things like 'That machine may be intelligent according to your definition, but it isn't according to mine.' So it is necessary to be clear about words and how we use them. The third chapter of the book attempts to introduce some clarity in this respect.

Another difficulty arises from the emotional overtones of some of the words we shall be using. We feel stroked, comforted and warmed when we are described as 'intelligent', 'human', 'creative' and 'alive', and correspondingly insulted and degraded if we are called 'mechanical', 'inhuman', 'predictable' or 'unintelligent'. So part of the discussion is bound to be coloured by the way we feel it affects our status as human beings. In fact, much of the anxiety that a notion such as artificial intelligence generates may come from the feeling that we do not want aspects of our humanity to be reproduced in machines.

A third difficulty arises from the vast size of the field and the question of how much detail it is necessary to supply. Many readers will already know how digital computers work, others will not. Some will know about neural nets and genetic algorithms, others not. To give sufficient detail of each of these areas (and others) to satisfy those who want to know more, while keeping the discussion sufficiently straightforward to satisfy those who are interested only in the bare bones – the bottom line – would make the book impossibly long and diffuse. On the other hand, to say nothing at all about them would invite the criticism that the version of computer science I had discussed was out of date.

The compromise I have attempted to strike amounts to a decision to give brief (and therefore painfully inadequate) accounts of the technical details, while adopting an overall strategy which allows me to present summaries of the state of these disciplines as pointers to what may happen later. Those wanting a more technical discussion may refer to some of the books listed in the bibliography. All these topics are discussed in Chapter 3: 'Problems with Words'. It is probably the most technical chapter of the book, and those prepared to be less precise about the use of words may choose to omit it at a first reading.

The fourth chapter, 'Computers, Theology and Science', sets computer science and artificial intelligence in the theology-science debate. That debate has usually been concerned with physics (especially cosmology) and biology (especially evolutionary theory). It has also often been characterized by defensiveness on the part of religious people which has led them to be dismissive of new ideas in science. Artificial intelligence is unlikely to fare any better. There will be those who will say that it is not a serious issue because machines cannot have souls. For them, that is where the matter will end. But whether or not computers ever develop the full-blown characteristics of human beings, the capacities of intelligent machines can only increase. This chapter attempts, amongst other things, to forestall theology from taking up a position from which there can only ever be an ignominious retreat.

In Chapter 5: 'God and the Mind Machine', we turn to the religious implications of all these questions about the human mind, consciousness and soul, and ask what we can learn about ourselves from the things which computers can and cannot, may and may not be able to do. In particular, is the now-popular thought that human minds are rather like programs running in machines, and so may be able to be reconstituted in heaven once the body dies, a helpful analogy? It returns to the famous android Commander Data and asks whether, if machines eventually develop consciousness and mind, there is any reason to deny them the same status in creation as other creatures whom God loves. It also makes the important point that, if a time should ever come when our lives are taken up to a considerable extent in dealings with androids, it may cease to be possible to conceive of our 'selves' in the absence of those androids (rather as today some people are equally wrapped up in their dogs and cats). And in that case we may reasonably ask whether, if androids cannot go to heaven, anything other than a severely cut-down version of each self can go either.

Chapter 6: 'The Universe, Everything and Life', draws the discussion together by setting it within the context of God's overall creative purpose.

The religious content of the book is to be found throughout the chapters, and especially in the last two. It is also perhaps evident in the unwritten text of the book, that the 'faith beyond faith' not only allows, but actually encourages us to explore possibilities raised by science without fear that in doing so we are in some sense entering into a pact with the devil. It is an essential part of my faith that God empowers his children to face all the possibilities that arise from the richness of his creation. In that spirit, we are called upon as Christians, in what has perhaps become the most famous split infinitive of all time, 'to boldly go where no one has gone before'.

1 – DREAM OR FANTASY?

What has artificial intelligence (AI) achieved so far? What may it achieve in the not-too-distant future? What are its long-term prospects? What if we allow our imaginations to stretch into the distant future?

Retrospect

Between 1950 and 1980, great progress was made in rudimentary artificial intelligence work, especially in the United States. That success raised the hopes of those working in the field, and some very optimistic and even wildly exaggerated claims were made about the seemingly imminent development of fully conscious, intelligent machines.

Much of what was said belonged more to the realm of science fiction than science fact. Which is not to say that it would remain forever fiction, but that it bore about as much (and probably a good deal less) resemblance to achievable aims and current technological competence as Jules Verne's 'Nautilus' in *Twenty Thousand Leagues Under the Sea* to the development of the nuclear submarine.[1]

No doubt some of the 'hype' was intended to do no more than secure research funding; no doubt some of it was offered in all seriousness. But AI research had largely moved away from grandiose projects, and the results, in such areas as voice and pattern recognition, have been impressive. One of the few ongoing projects still actively engaged in old-style AI research is based on entering vast amounts of data into a computer in a (possibly futile) attempt to supply it with an adequate knowledge-base to deal with the 'real world'.

The immediate future

All this may suggest that the problem has gone away. I doubt it. Whatever the technical difficulties we face, and however many decades or centuries they take to resolve (if they can be resolved), the brute fact is that we have identified an enthralling problem and that we do not as yet know how to solve it. The problem is, in essence, how to make an artificial human being.

The fact that there is a problem we cannot solve leads me to believe that we will chip away at it year by year until we do solve it. It is in the nature of the human being to do just that. Good problems are hard to find; many of the more tractable ones have already been solved; there are many who think we are approaching 'the end of physics'. All the clever people in the world are going to be looking for something to do. They will almost certainly descend on artificial intelligence in increasing numbers. And, as is so often the case, some of the funding they will require will come from actual or imagined military and security applications.

Nevertheless, we will probably spend longer on more modest challenges in the short term. Early AI was wildly ambitious. It massively underestimated the magnitude and conceptual complexity of the task it was undertaking, and duly found itself up against a wall when it had barely begun. There are more modest tasks.

IBM and others have already substantially solved the problem of speech recognition, but that too proved far more difficult than was originally appreciated. It is certainly possible that I shall dictate, rather than type, my next book, to a speech recognition machine. I could (just) have dictated this one.

Security systems are coming closer and closer to effective and reliable hand- and face-recognition, usually based upon neural net technology.

Computers can already read the printed word and speak it (albeit in a rather unattractive robotic voice, but that is once again merely a matter of technology development).

The time is not far off when most homes in the developed world will be connected to the Internet. Then banking, shopping, and most communicating that is currently done on paper, will be done electronically. At some stage education will begin to be home- and individual-based (although there is a social dimension to education which may prove more difficult to replace).

So these and other more immediately useful tasks will gradually familiarize us with (and, by the same token, immunize us against) the impact of the computer on daily life. Early in the new millennium, millions of people will be entirely at home with the idea that computers talk to them, understand their spoken commands, and undertake many routine domestic and business chores over and above those they already perform. What is more, many of these functions will be performed automatically. Bills will be paid not only by central computers operating direct debits, but by home computers responding to demands for payment.

Social changes do not take place overnight. Reports that make dire forecasts of how the world may be in fifty or a hundred years tend to generate alarm (or indifference) because they fall on a society that has not even begun to adjust to the changes they foresee. We feel that if we ignore these changes they will go away because they seem so utterly remote. We are tempted to feel about them much as we feel about dumping nuclear waste, overpopulation and pollution: that will be someone else's problem.

But that is to overlook the gradual way in which social changes come about. By the time we have grown used to talking computers that manage rudimentary aspects of our lives, read our children bedtime stories, and keep us informed of developments throughout the world on which we have expressed an interest, we will be ready, perhaps even eager, for the next stage of computer development. But what will it be?

(If you doubt this, if, for example, you say that you would never dream of allowing a computer to read your children a bedtime story, reflect on the fact that a hundred years ago there were no televisions to sit our children in front of, no book tapes for them to listen to as they go to sleep, and no computers for them to sit in front of for hours. The parents of a hundred years ago would certainly have denied that *they* would ever have allowed their children these 'inhuman frivolities'. Look at us now!)

Long-term prospects

All these developments (and more) may increase our sense that computers can perform tasks often associated with intelligence. If we are prepared to forego the requirement that the computer should also be aware of what it is doing, to all intents and purposes these computers will be intelligent. But that is the real issue.

The things that are possible in the immediate future lead by natural extensions of technology to the idea of more and more sophisticated machines that may evolve into androids. Such androids will no more have awareness than does the computer on my desk if they are simply extrapolations of current programming. 'More powerful' does not mean 'different'. Such technological achievements are of the same kind as improvements in the size, speed, economy and environmental friendliness of aircraft. More impressive, perhaps; more interesting, certainly. But not of a different *kind*.

One of the reasons why I refer repeatedly to chess-playing computers in this book is that they are good examples of a general phenomenon: that something we feel to be fundamentally human turns out to be programmable. Even as I write, in computer laboratories throughout the world, programmers are solving problems which I cannot solve, finding ways to do things which I naively think to be beyond automation. With almost every visit to a computer shop one finds oneself thinking things like 'Can they do that already?' and 'Can all this powerful equipment really be so cheap?' We used to think chess a uniquely human domain. We were wrong. About how many other supposedly unique human domains will we also be wrong?

Imaginary futures?

Most people have seen *Star Trek* and have encountered Commander Data; many have seen Arthur C. Clarke's and Stanley L. Kubrick's *2001: A Space Odyssey* with its haunting images of space stations rotating to the accompaniment of the 'Blue Danube' waltz. Almost everyone born in Britain since the Second World War knows of *Doctor Who* and remembers the nightmares caused by the Daleks and the Cybermen.

The computer called 'HAL' in *2001* has long been the epitome of the ultimate AI question: when does the machine servant become the master? The problem, for those who are not familiar with it, is simply this. HAL is in charge of life-support systems. He decides, for various reasons, that he can only complete the mission successfully by taking command of the vessel and, if necessary, eliminating the human crew. His powers are formidable (he can lip-read, for example) and he speaks in a rather lush, seductive voice. How much of this is possible?

Apart from the lip-reading, which is somewhat ambitious by any AI standards, I can see little here that is not technologically realizable. Control of life-support systems is already well within the bounds of current technology. So is speech (in a limited sense). The part of the story that requires more careful thought is the conceptual side. Can HAL think? Need HAL think?

I am not sure that HAL even needs to think. The fact that Clarke and Kubrick could imagine the kind of situation that arises in *2001* means that HAL's programmers could do the same. They might ask what would need to happen if the crew were all killed, or mutinied, or went mad. How could the computer be prepared to take over? Granted some scepticism about the sophistication of HAL's performance, but allowing for the way programs can already learn, and permitting software (program) technology a few more years of development, just how unbelievable does the scenario in *2001* seem? Or am I falling victim to the kind of over-optimism that derailed early forays into AI?

What are the prospects for intelligent, inside-out orientated computers in the medium and long term? And what are the social implications even of the basic kind of technological development that I have outlined?

To understand these questions we need to look in a little more detail at how a computer works.

Turing Machines

A Turing Machine is a theoretical model of a digital computer. A digital computer works by processing strings of zeros and ones. These are called 'bits' of information. Alan Turing was able to show that every such computer can be simulated by means of a very simple process. He suggested that because computers work by reading single on/off or zero/one instructions, this process can be reproduced if the computer reads such zero/one strings from a long tape one at a time and performs instructions corresponding to the sequence of instructions it receives from them. See Figure 1.

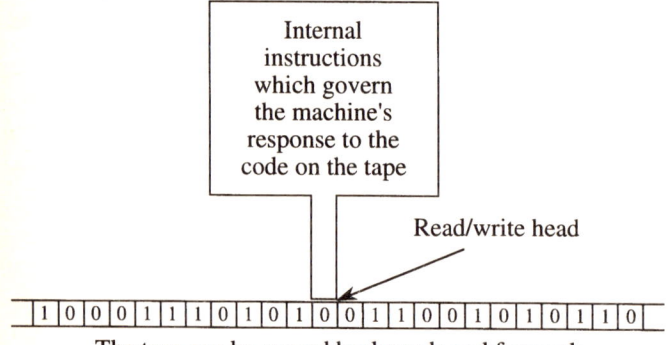

Figure 1 A Turing Machine.

Dream or Fantasy? · 13

Although the details of the way a Turing Machine works are fascinating, they are well documented in the literature,[2] and need only concern us in two essential respects.

First, that in order to know what to do with a tape, a Turing Machine must already have inside itself – embody – some rules about how to behave in response to what it reads on the tape, some operational principles. (For example, 'If the next number read is a zero, then ...'.) These principles determine what, granted its particular internal state, it should do next. The necessity for such internal software underlines the distribution of meaning over systems; it is just an illusion to suppose that the tape contains everything the Turing Machine needs to know.

The point at issue warrants further illustration. We are all familiar with the idea of a code or cipher. I may, for example, agree with a friend that if I insert an advertisement in the personal column of *The Times* saying 'To Jemma: Uncle Albert died on Friday 28th July 1995 at 1545. Please ring Paul.', it may have nothing whatever to do with Jemma, Uncle Albert, death or Paul. It can mean quite literally anything. Its function is solely to act as a trigger – a key – to set in motion some desired sequence of events. In other words, there is no fixed relationship between a message and the consequences, interpretation or meaning of that message.[3]

In the case of a Turing Machine, the equivalent of the message published in *The Times* is a string of zeros and ones on a tape which it reads one digit at a time. The machine can only know what to do with, say, 0101101000101000101111100 if it is prepared in advance for what the sequence 'means'. The word 'means', however, is really very misleading; all the computer has is a set of instructions which tell it how to respond to the input stimulus of such a binary sequence. Most of our attention tends to be directed towards the input, and little or none to the interpreting system which turns that input from a meaningless sequence of binary digits into an instruction which makes the machine do something.

Second, there is no question of the Turing Machine 'understanding' what it is doing (however intelligent it may appear, however well it may play chess, or however well it may write poetry). The Turing Machine may therefore *seem* to demonstrate that a computer cannot be conscious or engaged with. Whatever it does seems purely automatic. There would be no role for consciousness even if it were capable of it.

If understood this way, the Turing Machine is an exact analogy of John Searle's famous Chinese Room argument. A man in a room is passed Chinese ideograms through a window. He has no knowledge of Chinese, but he has a set of rules telling him which symbols to pass back. Searle argues that however well he persuades those outside the room that he understands Chinese, he does not in fact do so.

Searle's test is intended to show that a symbol-processing computer *cannot* understand. It actually shows only that the appearance of intelligence *need* not be accompanied by understanding, which just puts us back in the dilemma the Chinese Room argument is intended to resolve: 'this computer seems to be behaving intelligently, but does it understand? Is it aware?' That something can simulate intelligence without understanding – by just following rules – is not sufficient to prove that it cannot understand or is not aware. Otherwise, the test could be used to draw the same conclusions from human intelligent behaviour.

Long before Searle, Alan Turing proposed his famous Turing Test which any computer would need to pass if it were to be considered intelligent: that a human being, communicating with that program through a terminal connection, would be unable to distinguish it from a human being. This Turing Test has celebrated status in the literature. But it is similarly bogus. It is the inadequacy of this test that Searle's Chinese Room argument demonstrates (not the impossibility of computer understanding). The test confuses the ability to appear to be something with being that something. A program which manages to persuade us that it is intelligent does not thereby qualify to be considered human. A simulator

Dream or Fantasy? · 15

which persuades us that it is a jumbo jet in every detail of its operation still lacks one vital quality: it is useless when we want to fly to New York.[4]

Stored knowledge

There are, in broad terms, two kinds of knowledge: knowledge that is based in the make-up of a structure and knowledge that is learned. Evolution gives bodies the former kind of knowledge; life provides the second. Knowledge that is stored in bodies from birth (and therefore conveyed somehow by genes) we may call *embodied* knowledge; knowledge that is acquired we may call *learned* knowledge.

EMBODIED KNOWLEDGE

When we turn the ignition key in a car and the engine bursts into life – a process we take for granted until it fails to work – we are making use of an important feature of any tool or machine, that it somehow carries or stores within itself the know-how required to make it work. This inbuilt know-how is embodied knowledge, knowledge that is held in the way something is made. It is not knowledge that is conscious, or the kind of knowledge we need to express in language before it can be used. It is just an inbuilt capacity to do certain things, and often to do them extremely well.

Living things, like machines, rely upon this kind of knowledge. We no more tell our eyes how to see than our stomachs how to digest or our hearts how to beat. Our bodies, as a result of a long process of evolution, know how to do all these things. Babies are born curious; they do not learn curiosity. They are born with an inside-out world. The inside-out-ness of living organisms is hard-wired (embodied) from birth. We take these abilities for granted until they go wrong, and because we scarcely understand how they arose or how they work, we are often at a loss to put things right when and if they go wrong (but then most of us are similarly stuck when the car breaks down).

When we switch on a computer, we probably know as little about what goes on inside it as we know about the workings of a motor car or our own bodies. The disk drive whirs, the screen lights up, we press buttons or click icons to load programs, and away we go (writing a book, doing our accounts or playing a game). The computer embodies 'how to get started'.

Unlike a motor car, where the input of the driver, although continuous, is limited to a small number of controls, our input to a computer can be much more elaborate using keyboard, mouse, modem, scanner or joystick. But the machinery of a computer requires something in addition to its design and the input of the user because the computer hardware (the metal case, the disk drive, the memory, central processing unit and screen) only has built into it part of what it needs in order to work, whereas the hardware of a motor car embodies everything, lacking only the driver's guidance.

It is helpful to consider what a motor car really is because that helps us to understand what the important qualities of a machine are. It has no program (leaving aside smart cars with computers to tell you how much fuel they're using – most of us would rather not know), but it does have an open quality which means that, by moving certain levers, pressing certain buttons and turning the wheel, we can direct it as we choose. This open quality (that the life of the car is not completely fixed) is one of its most important features. A car which could only travel one route at one speed to one destination would be of little use except as a convenient way to get to work. Instead we supply the car with controls, open ends which we can work to direct it in different ways. But the car is in the end no more than a car and it cannot change what it is. A computer is different.

LEARNED KNOWLEDGE

Human beings learn to speak in their native languages, but there are no inbuilt differences in their bodies to accommodate different languages. This knowledge is learned and stored. In the case of the computer, it was John von

Neumann's realization that it is possible to superimpose different programs on the same basic hardware structure that set off the computer revolution.

By changing its non-essential programs, a computer can be made into a game machine, a word processor, a calculator, a database, a spreadsheet, a graphic design studio, a music centre, a means to communicate with the rest of the world. At no point does the machine itself change. The hardware (with a few additions here and there) remains the same, but the program running in the hardware makes electric currents flow in different ways to produce changes of patterns on a screen, changes of movement in a robot, or changes of signal in a communication channel. This ability to change its nature means that its flexibility is awesome; no tool like it has ever existed on the planet before. The only other objects in the known universe that have this flexibility are living organisms, and most of them other than human beings are far more restricted in what they can do than the modern computer.

A new Babel?

Humankind has made the computer in its own image.

Some would argue, and others fear, that we may have unleashed something that will overwhelm us, and that the only sensible response to these offspring is to destroy them while we still can. Others are simply anxious about our growing dependency upon computer technology (although the same could be said for electricity). For who is to say whether or not computers will one day prove more flexible and adaptable and durable than even human beings themselves? It is not only religiously minded people who have felt inclined to say this.

There are parallels in both the story of the Tower of Babel in the Bible (Gen. 11:1-9), where human beings became so clever that they aspired to become as gods, and in the story of Kronos in Greek mythology, who routinely ate his children for fear that they would rebel against him.

Of course, if we believe that something is impossible, we need not worry about what would follow if it happened. But there is an obvious problem with believing that it is impossible for brute matter to generate inside-out-ness in the shape of minds: brute matter has already done so, because it has given rise to human beings and other creatures. My worry, if I have one, is almost the exact opposite: that while most of the human race believe the generation of artificial intelligence and life to be impossible, the minority will actually achieve it. This problem, as I shall say often enough, just will not go away.

Of course, computers are so powerful, fast and flexible that it is easy to be carried away with enthusiasm for what they can at present or may one day be able to do. Those who write programs have often fallen foul of the temptation to make predictions about computer advances that have not materialized. The power of present computers seems to them to point in the direction of computers with endlessly increasing power. It seems to them that they will employ more and more sophisticated programs that will one day be capable of almost anything, including all the thoughts that a human being can think. Is this claim justifiable?

To see one flaw in the argument, consider the motor car again. An alien, watching a car negotiate a busy street, weaving in and out among other cars, might be tempted to think that the car was clever or intelligent, because it responds appropriately to changes in circumstances that it could not really have anticipated. But we know that all the intelligence lies in the driver and (in a different way) in those who designed and made the car to be open to the driver's control and therefore flexible in use. By combining a clever, open design which allows it to be controlled, with the skills of a driver, the motor car comes to look intelligent. But this is an illusion: the motor car simply transmits or passes on or bears witness to the cleverness of those who made and drive it.

On the other hand, the program running in a computer is not directly under the control of a driver in the way that a motor car is. An alien who could communicate with such a computer – a computer which replied to its questions and gave adequate responses to them – might be convinced that the computer was intelligent. Here the distinction between human intelligence and machine intelligence has been blurred. Once a machine can stand alone, without direct human input, and persuade us that it is intelligent, it has moved into a new sphere of sophistication. (That no such computer currently exists is not in itself a reason to deny that it may do so one day. The question of that possibility is very much one of the questions this book addresses.)

If a computer appears to hold a conversation with me, is the intelligence it displays no more than a transmission of the intelligence of its designers and programmers, even if their efforts are buried deep inside it and in the past? And if so, is a computer really any different from any other machine? Would an alien watching a computer have a conversation with a human being be right to think the machine intelligent? (Would the alien even be right to think the human intelligent, given that it was talking to a machine?)

Models for the mind

Although many of the things we hold to be characteristic of human beings are certainly things computers can do better than we can (or will one day do better than any human), the way in which some of us have come to think of human beings as programs running in machines is based upon a deep mistake produced by the absence of an adequate alternative. The sheer clarity of the computer model beguiles us into turning it upside down. There are things – things whose rules we can express directly and clearly without ambiguity (like the rules of chess) – which are naturally well suited to analysis in terms of computer programs. So there are many things where the computer model is entirely appropriate. But instead of thinking only of things of such precision in terms of computers, we have tended to assume that the simplicity and beauty of the computer model must mean that everything can be modelled using it.

This tendency to allow models to be applied inappropriately can be called *model inversion*: instead of thinking of the world using models we know to be inadequate for want of anything better, we allow those inadequate models to reshape the way we think of the world.

Because computers are everywhere, and because the notion of a program has become so familiar, we tend to use it to describe almost everything. We think of brains and minds in terms of programs for want of any other way in which to think of them.

Model inversion is a serious problem in philosophy. It appears in many forms. For example, some writers believe that the way we think of the world governs the nature of the world. This is pure model inversion. It allows what we are capable of thinking to dictate what is there, as if the world cared what we are capable of thinking and politely shaped itself accordingly. Obviously we are limited by our knowledge, for to know what we did not know (that is, where our knowledge falls short) would enable us to alter and improve it, but the idea that what we know (or how we know) dictates what is actually there is frankly ridiculous. Instead we need to be aware that the way the world is, almost always proves to be far more subtle and complex than our most advanced theories: our ability to know does not constrict the world's ability to be. (After all, the world came first.)

This may seem something of a digression. Why does it matter in the context of the present discussion that we allow our models to shape the way we think of the world? What alternative is there?

It is certainly true that we always think of the world using some kind of model, something that is not really like the world or the way the world is but which we use to help us along. Apart from our sight, for example, there are no red or blue objects; we say there are red or blue things because of the kinds of eyes and brains we have.

It is the same with language. We single out certain collections of atoms as things because they are important in our lives; large red moving objects on stretches of black tarmac are to be avoided if we do not wish to go under a bus. But it is perfectly possible to imagine a creature with different senses or with a much larger or much smaller body for whom the bus would not even be a part of the world. The kinds of creatures we are shape the kinds of world we think we live in. But we should not imagine that the world as it is, quite independently of our thinking and knowing of it, is bound to conform to our way of thinking and knowing it.

The trouble is that certain models tend to control the way we think, and although it is impossible not to use models at all, it is extremely important to have lots of different models to choose from. And it is important to remember that, rather like our analogies, our models sooner or later break down. They may even break down right at the beginning by being unsuited to a particular set of events or circumstances. If we keep these warnings in mind, no model ever takes complete control of the way we think. And that is especially true of computer ideas as models for the way we think. (It is also true of mechanical ideas of cause and effect which are not directly of concern here but which plague discussions of, for example, the ways in which God may or may not act in the world.)

In the absence of different ways to think of things, we fall back on familiar but possibly inappropriate ways to think of them. For how could we think of things using ideas we cannot understand or have not yet invented? There is no answer to that until we have some different ideas to choose from. And even when there are alternatives, a model such as the computer – when it swamps our intellectual universe, forcing us to think of everything in its terms – leaves no room for those other ways to think. We become victims of a kind of tyranny. Hence the quotation from Joseph Weizenbaum with which the book begins; we need more resources to call upon.

At this point those persuaded that computers are (or will become) all-powerful play a very dangerous trump card. They ask, on the basis of all that computers have already achieved, whether it is reasonable to doubt that they will one day be able to do (more or less) everything. (Sir John Eccles, the famous neurophysiologist, has called this *promissory materialism*.) This is rather like the famous God of the gaps argument in science, where, as science reduced the gaps in human knowledge, theological accounts increasingly seemed redundant, and were then squeezed into the remaining small spaces. Many of the things we were once tempted to say computers cannot do they now can do, so we are more cautious about what we say they may not be able to do. But there is a flaw in this argument.

Consider the world record for the mile. It was once over four minutes; Sir Roger Bannister broke that barrier, and soon many others did too. How fast will we one day be able to run a mile? There is clearly a limit. The fact that the record goes down (and will keep going down) does not mean that it will go on down to zero. Instead we measure times more and more accurately. Where once a tenth of a second constituted a new record, it became a hundredth; one day it will perhaps be a thousandth (as with racing cars). But the mile will never be run as fast as the half mile is run. There are limits. So the argument that runs 'We have always progressed, so everything will one day be achieved, every limit will one day be exceeded' is philosophically naive. We will just progress more and more slowly and approach an ultimate limit more and more gradually. The question is, where is that limit? Is it above or below the point at which computers can do everything human beings can do?

In recent years we have witnessed almost unbelievable improvements in the speed and quality of computer technology. The changes can be measured in terms of the home computer market (although similar changes have occurred at supercomputer levels).

Similarly impressive progress has been made in software (program) development. We now take for granted word processing that automatically paginates, with footnotes, formatting, laser printing, faxing and so forth. The word processors which were on the market in the mid-1980s were crude and clumsy by comparison. Graphics capability has altered even more. Millions of colours can be displayed; moving images of two-dimensional representations of three-dimensional objects are within the compass of relatively modest home computers. We can manipulate, edit, design and integrate interactive multimedia systems in our own front rooms, and access databases the world over via the information superhighway.

But smaller, faster, bigger and better computers do not necessarily imply a future of limitless possibilities. We remain on the steeper parts of the human learning-curve, and there is no telling when we may reach some technical limit. Hardware limits may be caused by, say, our inability to miniaturize any further or the problems of dissipating the heat generated by the central processing unit. Software problems have already destroyed the early optimism of the artificial intelligence community. What seemed like problems that were on the verge of solution in the seventies and early eighties have now been perceived to be profoundly intractable. The Frame Problem is perhaps the most famous and difficult of them.

The Frame Problem

Human beings are wonderfully good at dealing with language in the everyday world. The vagueness that is such a problem over words like 'machine' and 'life' does not usually worry us at all. It is one of the strengths of language that it allows such flexibility.

It turns out that trying to program a computer to cope with such vagueness is unbelievably difficult (so difficult that many researchers have actually given up trying). Computers have to be told absolutely everything; nothing at all can be assumed. But in the human case almost everything (in normal conversation) can be assumed. And this skill is not restricted to a language we speak fluently such as our mother tongue.

Even a simple competence in French or German enjoys and relies on this skill. The fact that I learned how to say 'table' in German in a room where all the tables were made of brown-painted metal does not prevent me from using the word in millions of other contexts without difficulty. Yet a computer simply could not do this because it would need to be told every single instance in which the word 'table' in any language could properly be used. And that is a virtually impossible task.

What I have just described in general terms – the difficulty of telling a computer everything it needs to know if it is to be able to 'assume' the kind of background knowledge we take for granted when we use a word such as 'table' our language – is called the Frame Problem in artificial intelligence. It is called that because all knowing arises within a 'frame' or against a background that human beings take for granted and that computers need to be told about.

Even putting the matter this way makes it sound deceptively simple. 'Telling' a computer something involves giving it a rule or a set of rules which are expressed in some formal way. But many of the things we assume are very difficult to express in rules, and some are probably impossible to express that way because the rules themselves are only understandable if we have already mastered how language works. How much, for example, would you need to be told in order to be able to use a word like 'table' completely fluently and accurately and appropriately? It is just hopeless to try to list all the things. Every time we enter a new room in which there is an object we would call a table, we encounter a situation where 'table' can be used which we have never met before, so (since no rule can anticipate all such situations) we cannot use the word by following explicit rules. Somehow human beings know what a table is, even if they do not quite know how to say so. (We are confronted by a slippery bar of soap – see p. 45.)

All this talk of language may seem very abstract. An example from everyday life may make more sense, and it certainly illustrates the Frame Problem very clearly.

AN ILLUSTRATION

The problem of the everyday and how we make sense of everyday events may be illustrated with an anecdote about my own relationship with a humble kitchen kettle.

When I walk around my house I see various objects located more or less where they were when last I saw them (it all depends on how active the children have been in the interim). I am making a cup of coffee. The kettle is familiar; the coffee-grinder is less so because until recently we always drank instant; the cups are vaguely familiar, but we have so many that I am not totally aware of what each of them is like; I know where the milk should be (the triumph of hope over experience).

Let us now consider my response to the kettle in a little more detail. An observer watching me fill it and switch it on will not see that I actually rather dislike it as a kettle. I bought it myself, but it was not one of my better purchases. It is badly designed in several respects: the water filter it came with prevents the boiling water from coming out of the spout, and it promptly comes out round the sides, all over the worktop rather than into the coffee pot; it is described as 'cordless', but it fits its base so badly in order to achieve this inaccurate accolade that it would be better were it otherwise; the lid fits badly, and has handy little grooves in it that fill with water when you fill it and which then decant water (unboiled of course) into the coffee pot when you tip the kettle up. All these thoughts form the background to my inside-out perception of the kettle, together with vaguer memories of the day I bought it. I pondered ruefully at the time that the entire human race seemed incapable of designing a simple kettle that fitted even the most basic requirements of convenience and safety. So these inner reflections go on. Of course, neither I nor you the reader are greatly interested in them, but the fact that they are there (considerably more to the fore of my consciousness as a result of writing all this down than they are when I use the kettle several times each day) colours my perception of the kettle, and therefore my perception of myself as reflected in and by my use of the kettle.

A MINIMIZATION STRATEGY

This kind of stream of thought makes up the background – or the frame – of our everyday consciousness. Even if we are not very interested in such reflections (and who would be?), we are aware throughout our lives of the fact that things are seldom merely things. We connect the things we encounter and that happen to us with countless other thoughts and experiences.

A typical response by someone with high hopes for machine intelligence will be to say that this just is not fair. To expect a computer to have such feelings about a kettle would be absurd because it is far too sophisticated a level of awareness. The AI researcher will concede that we are nowhere near to solving this problem. Instead, AI reduces its expectations. It moves from its early optimistic predictions of machines that will one day enjoy all the qualities of the human (including all the conscious and background thoughts associated with such everyday occurrences as boiling water in a kettle). They are cut down or minimized. AI ignores the complex associations of an action and settles for the mere performance of the action (where that is attainable). But this is a minimization strategy and a Sliding Definition Ploy (to which I will refer again at the beginning of Chapter 3). By replacing the rich associations of the performance of an action with its mere performance, AI potentially (if not deliberately) deceives us into believing that a machine which seems to be human, on the basis of the range and subtlety of its actions, must therefore be human. (This is *positivism* – the conviction that only the things which we can see and measure are really there. Such an assumption is also built into the Turing Test.)

A naive theory of language would say that to know how to use the word kettle and to know what it means involves no more than knowing what objects are called kettles and what things are rightly said of kettles. My rather frivolous example shows this to be grossly over simple. To know what 'I' mean by 'kettle' you have first to understand the frame in which my much-maligned kitchen kettle exists, and that involves knowing far more than how to use the word.

If, as seems likely, you have started to have serious doubts about my sanity as a result of all these complaints about a humble and no doubt innocent kettle, explaining what 'kettle' means 'to me' may involve you in an explanation of my rather warped (as you may think) psychology. Understanding this one word becomes a massive undertaking.

The *qualia* problem

Qualia are qualities, felt experiences. We see colours; we hear sounds; we touch textures. These are *qualia*, the qualities of the world as we experience it. *Qualia* are properties of the inside-out world that cannot be seen from-outside-looking-in. You may very well somehow see certain parts of my brain operating in ways that suggest that I am seeing, hearing, smelling something, but that knowledge will neither allow you to tell what I am seeing, nor how I am seeing it and what impact the experience is having on me. From-outside-looking-in you can only guess or surmise that if we are both looking at the same yellow scrambled egg, it tastes to me as it tastes to you. It is much harder to imagine what it might taste like to a bat, and even more difficult to surmise what it might taste like to a cockroach (in Thomas Nagel's terms[5]).

Most of us would concede that scrambled egg does taste like something to another human being, that it probably tastes like something to a bat, and that it may taste like something to a cockroach. But could it ever taste like something to a computer?

Certainly we can devise a computer with sensors capable of detecting scrambled egg, and certainly we can program the computer to display 'This is scrambled egg' on the screen. We can almost certainly enable the computer to distinguish between different kinds of egg cooked to different recipes, and we can certainly program it to comment on those differences in pertinent and precise ways.

But what does it mean to say that the computer tastes the scrambled egg? This is the *qualia* problem. It is a source of great embarrassment to the AI community, and their response is either to deny that the problem exists or to argue it away. These are both minimization strategies. That a computer can be designed with circuits capable of identifying a substance as scrambled egg is not in dispute. That identifying that substance as scrambled egg is *the same thing* as tasting it, is very much in dispute, and is in fact the crux of the whole AI debate.

It may seem from our own experience that we identify what a substance is by, for example, tasting it. 'I know this is scrambled egg because it tastes like scrambled egg.' But various experiments on the autonomic nervous systems, and even more dramatically on split-brain patients, have shown that we are capable of identifying things correctly *without* ever being conscious of their *qualia* (their taste, their colour, etc.).[6]

The 'arguing away' is an extension of a philosophical refusal to acknowledge that conscious experiences as we experience them are irreducible to biology, chemistry or physics. It is an example of taking one-kind-of-stuff views to an extreme, and denying the primacy of the very experiences necessary to action that denial. In other words, the meaning of 'This sentence doesn't really mean anything not reducible to physics and chemistry' is supposed to be something that is purely a matter of physics and chemistry. But this is just like sawing off the branch you are sitting on. The point of making such an assertion is to communicate, not to initiate a chemical reaction. And it is persons who communicate, not bodies.

The meaning problem

The meaning problem is the linguistic extension of the *qualia* problem. We imagine that meaning is contained in words. As you read these pages, my meaning (almost certainly distorted to some extent) comes over to you through the words which I am now typing. It seems clear therefore, since the words are the only connection between me as author and you as reader, that the words contain the meaning.

This is just not true. Suppose I suddenly type something like %^+4)9πio weoi 90304w rew. You suspect a misprint. 'This doesn't make sense.' You write rude letters to SPCK about their poor proof-reading, or ruder ones to me about not making my meaning clear. I write back and ask why you have a problem. My meaning is perfectly clear – if the words contain the meaning.

'Ah, but your words are not written in English' you reply, triumphantly. And that is my point. What does this 'not written in English' actually involve? It involves the words I write drawing (or, rather, failing to draw) upon a set of meanings that reside not in my words but in the English language as a whole (and therefore in all the people who use the English language).

Were we to come across the written remains of a lost civilization in a tongue no living human could speak, we might easily deduce that something was being said by those words. Their pattern, in particular the way certain things were repeated, would show that they were neither purely natural nor purely random. But we might have the greatest difficulty in deciphering them.

So meaning does not reside in words. It resides in communities which use words. The connection with the Frame Problem and the *qualia* problem is now easier to see: how much of the understanding of a community do we need to absorb before we can make sense of its written and spoken language? How much of that communal meaning would a computer need to be given if it were to be said to understand a language? For, without meaning, all the computer is doing is manipulating marks and symbols. It understands nothing; we supply the meaning (just as you are supplying the meaning that brings my printed words back to life by drawing upon your mastery of English). Without the meaning we bring to bear to make sense of the rules they follow, the symbols and words computers produce mean no more than the marks you see before you on the page unless you, or someone like you, reads them.

Computer output can be made to seem intelligent by creatures who are themselves intelligent because they, not the computer, supply the frame within which what would otherwise be mere marks are turned into a means of communication.

So far, then, we have seen that one problem which distorts discussions of machine intelligence is that human beings habitually bring their own meanings to bear upon things they encounter in the world. Because computers, programmed by human beings, generate output that is shaped by human beings (indirectly), when other human beings encounter that output they tend to impute meaning to it, and intelligence to the device which produced it, forgetting that it is they, not the computers, which are responsible for perceiving such meanings as may be apparent.

This deficiency can be seen if we ask an important question about thought.

The thought problem

If there were only computers in the universe, would there be any thoughts? Not unless computers are developed which develop inside-out worlds.

Even if a computer were to be created which passed a rigorous Turing Test, that machine need not be considered to think or to have thoughts. Instead we should say that it creates sentences which induce thoughts in human beings or other organisms capable of understanding a language or a symbol.

The fundamental thing that separates living organisms from computers as we know them is that they are aware. They have a sense of being there that extends beyond being merely users and producers of sentences in a language and actions in a world. They not only live and breathe and eat and act and wake and sleep; they register being alive in a way that exceeds merely stating that they are alive.

Very few artificial intelligence (AI) scientists really acknowledge the force of this problem. Instead they turn the problem round by saying that what 'I' am aware of is irrelevant if others cannot know that I am aware of it. In other words, they say that my awareness of being alive is of no consequence unless some other creature can, by interrogating me rather like someone administering a Turing Test, deduce that I am aware of being alive. To put it yet another way, they claim that, failing success in this project, my awareness of being alive (and yours) are not really 'things in the world to be taken seriously'.

This is completely absurd, of course; there is nothing more primitive, more basic, more absolutely primal than our awareness of being alive, of having experiences, of being in some sense centres of experience. An argument to the effect that my awareness of self is somehow 'really' just something else is so ridiculous that only someone concerned to assert it for some ulterior motive could conceivably advocate it (such as that it is a major source of embarrassment to the AI fraternity). My awareness of being alive is the most immediate and indisputable fact of all facts; without that, all else is as nothing. Descartes, with his 'I think, therefore I am' was (at least in this respect) absolutely right (although perhaps for different reasons from those he gives and in a way that serves purposes he could not have envisaged).

The brain problem

Perhaps the most obvious reason to doubt the eventual success of strong AI arises from the fundamental differences between brains and computers.

Brains are nothing like digital computers; their structure is more like that of a neural net. But no neural net has come anywhere near even to the degree of structural complexity of a brain (where every one of tens of millions of neurones may be connected to as many as ten thousand others), to say nothing of the embodied knowledge base of the brain.

The capacity for inside-out-ness of human beings (and all other higher organisms) is hard-wired; we and they are born with it. So not only do we need many connections; we also need specific, learned, embodied connections between our neurones. And the degree of complexity involved is simply vast, well beyond the capacity of even the most advanced and ambitious neural net research program.

Will time erode this deficiency? Certainly, but, as Hubert Dreyfus once so scathingly put it, it is unlikely that neural nets will have developed the sophistication of a slug in one hundred years. Maureen Caudill puts it like this.

> What are the limits of performance of a neural network? Frankly, no one knows.... But ... I can very confidently assert that no theory will ever prove that neural networks are incapable of being intelligent. Why? Because we have at least one well-accepted example of an intelligent and capable neural network mind: that of a human being.[7]

This is true provided, of course, that one accepts that the brain of a human being is nothing more than a very complicated neural net. Yet there also seems to be another kind of difficulty, which lies in the way that the brain is able to draw together and focus wildly different sets of experience in order to create new thoughts and ideas. I call this its ability to *correlate* its activity.

COLLISIONS AND CORRELATIONS

The temptation to regard the brain as a neural net leads us to try to model it using that concept, because we lack another way to think of it. In this section I shall outline a possible problem with such a model which relies upon the distinction between a *collision* and a *correlation*.

The best way to illustrate the difference is to consider a snooker ball striking a pack of reds and sending them in all directions. This is a collision; one cause produces many effects which may be so complex that we are almost incapable of keeping track of them. Science is good at dealing with collisions; they are the basis of all its experiments and machines. We control a few inputs (such as switches), and we arrange the apparatus so that everything else follows from them. The switches, so to speak, act like the colliding ball to initiate further changes in the system. Turning the ignition key in a car; switching on the television or computer; firing a gun; aiming a snooker ball at another; dropping a pebble into a pond – these are all collisions in this sense. See Figure 2.

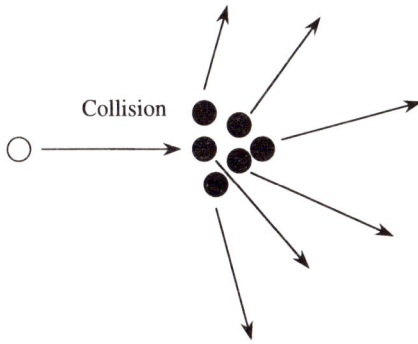

Figure 2 A collision.

Correlations are the exact opposites of collisions. The snooker balls which are spreading out and bumping into one another are correlated because they share a common history which originated in the collision, and if we could instantaneously reverse all their motions perfectly and exactly, they would move back to assume their original positions again, and project the cue ball back to the tip of the cue. A correlation that reflects the way things originated in the past may, by analogy with this process, also be capable of producing a single effect in the future. Rather than reversing motion (or even reversing time), we can imagine that countless apparently independent and unconnected factors (which we could call multiple causes) might in fact be correlated into a convergent pattern of activity that gives rise to one single future event. Here our understanding of their significance for one another could only occur once we had observed that future event and seen how important (and perhaps improbable) it was. We might say, 'What a coincidence that all the necessary factors combined to produce that effect', and be led to look for what would prove to be a complex set of coordinated, correlated causes. See Figure 3.

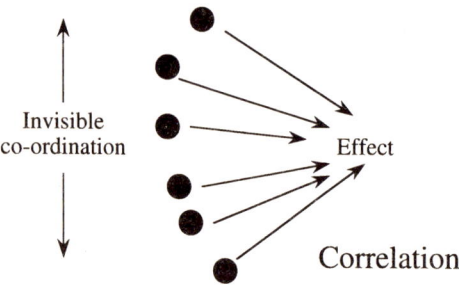

Figure 3 A correlation.

I wish to suggest that what makes mechanical, collision-like descriptions and models of the brain so inappropriate, is that the brain actually works by generating exactly these kinds of correlations, and that its workings are only ever apparent from its results. By some process of which we are at present in almost total ignorance, the brain somehow manages to coordinate large numbers of its countless parts (modules) in ways that produce non-localized things which that same brain, from-inside-out, experiences as thoughts. The brain, in short, produces thoughts by generating and controlling correlated neural activity.

In practice, the brain is prompted into such action in most circumstances by collision-like events such as incident stimuli which prompt it to respond. The stimuli set in train a host of brain activity which, sometimes almost immediately and sometimes over a much longer period of time (even days, months or years), produces correlated activity that gives rise to a thought or string of thoughts.

Such strings of thoughts are commonly mistaken for a largely imaginary process called 'thinking' which human beings believe themselves to engage in, but which in fact consists of saying to oneself, 'I must think about that' and then waiting passively for an answer to spring to mind. But strings of thoughts need exhibit no logical or rational connection with one another, even though they are produced by complex correlated brain activity in response to some perceived (possibly also very complex) problem.

For example, this book when finished will be about forty thousand words long. Even I, its author, will not be able to hold every one of its ideas in my conscious mind at once. Yet all its sentences and ideas arose from my brain, which produced, over a period of time, strings of thoughts which as often as not came as surprises even to me. The coordinating activity which renders those hundreds of thoughts and thousands of sentences coherent must lie somewhere within areas of my brain over which I have minimal control and of which I have scarcely any consciousness. What I can do and have done is to 'aim' my brain in certain directions by using my consciousness as a way of selecting certain kinds of areas of interest from among the many which each of us could study and write about. My conscious decision to write this book set in train a process which what we sometimes call the unconscious mind worked on even when I was doing something else. (Indeed, it is often the case that I find my most fertile ideas pop into my head immediately after a period of time when I have been singly absorbed in something else with nothing to do with the topic I am writing about. It is almost as if by ceasing to try to solve a problem using the ham-fisted attempts at thought that are within my control, I free the rest of my brain to solve it far more efficiently.)

In the *Zettel*, Ludwig Wittgenstein writes (but using 'correlated' in its more usual sense):

> No supposition seems to me more natural than that there is *no* process in the brain correlated with ... thinking. I mean this: if I talk or write there is, I assume, a system of impulses going out from my brain ... But why should the *system* continue further in the direction of the centre? *Why should this order not proceed, so to speak, out of chaos?*[8] [My emphases]

I think this is absolutely correct; a simply brilliant insight. Wittgenstein is not advocating some kind of mysticism, of course, as if nothing is going on and thoughts proceed from a little man (homunculus) in the brain; he is saying that the ordered neurological processes responsible for a conscious thought may be the first systematic realization in the brain of anything remotely like that thought: the system need not go deeper. 'The one thing that I cannot prepare myself for is the next thought that I am going to think.'[9]

If there is any mileage at all in this conjecture, it means that the responses of brains to stimulation, for example in the production of thoughts, are unique to each individual brain, and that brain activity is so unlike computer activity that all attempts to model it using computer processes are made to appear pitifully short-sighted, futile and misconceived. See Figure 4 for a diagrammatic representation of this.

Of course, from-outside-looking-in all we could see if we were monitoring the brain would be a jumble of neurological processes, almost certainly lacking any kind of intelligibility or pattern, and least of all any coordinated correlation. Then, suddenly, the subject would (perhaps) make a statement: 'the answer is 42!' Somehow, the chaos below thought would have produced an intelligible effect from a multi-causal but impenetrable jumble of correlated processes. (And, precisely because they are impenetrable, it is not always easy to see what

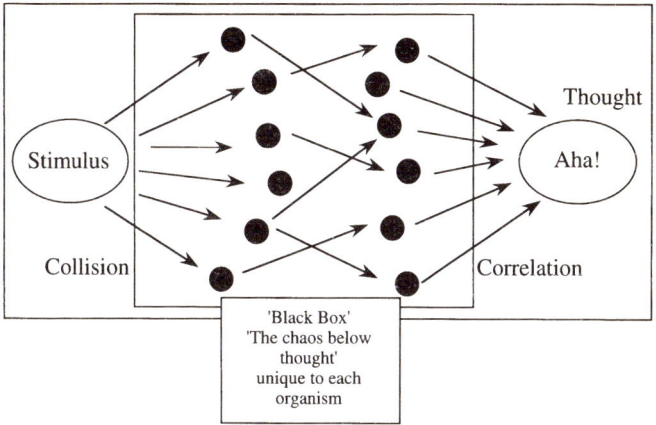

Figure 4 Each brain is unique.

has gone wrong if the answer the subject gives is incorrect, and difficult to see what else we can do if – as is so often the case – the awaited answer just will not 'come'.)

From-inside-looking-out, the chaos below thought that is visible to the scientist observing it from-outside-looking-in is completely invisible. Other things are occupying the subject's conscious mind that may have nothing to do with the problem to which '42' is eventually given as the answer. As far as the subject is concerned, the chaos below thought runs in the background, sometimes producing good ideas, sometimes generating nonsense, sometimes offering suggestions that lead the subject into trouble, and as often as not producing nothing useful at all. Mostly the subject is simply amazed: 'However did I come to think of that?' she asks herself.

To sum up: mind – as the world-orientation of body, its inside-looking-out-ness – is oblivious to the chaos of neurological activity essential to its existence. The brain must work properly if the mind is to exist and be healthy. But minds are never directly aware of the brain upon which they depend. Minds, for example, *taste*; brain circuits *fire*.

Minds arise from the workings of brains in terms of qualities (*qualia*) and meanings. They are stimulated by the world in which they live and they rely upon the capacity of the brain/mind system to solve problems by focusing widely spread sets of experiences. The effects of these correlations are experienced as thoughts.

All this requires us to consider in more detail the connections between minds, brains and consciousness.

2 – MINDS, BRAINS AND CONSCIOUSNESS

It is not possible to discuss mind and body without talking about dualism and monism. I shall call mind/body dualism a 'two-kinds-of-stuff' view (the view that there is mental stuff and physical stuff and that they are different kinds of stuff), and mind/body monism a 'one-kind-of-stuff' view (the view that mental and physical realities are to be explained in terms of one kind of stuff). The view I shall argue for involves having only one kind of stuff, but different mental and physical realities, neither of which can be described adequately solely in terms of the other.

Two-kinds-of-stuff views

For much of human history we have believed that we are really two different kinds of thing: bodies and souls. The bodies are made of flesh and blood (matter), but the souls are not. Plato, for example, believed that souls were somehow trapped in their bodies during their lives, having existed before in the perfect world of ideas, and that after death they go back there.

Those who believe that there are two kinds of stuff (matter-stuff and soul-stuff or physical stuff and mental stuff) have always tended to look down on the physical as inferior to the mental or spiritual, and on earthly life as no more than a prologue to eternal life in the realm of perfection. This obviously connects with religious ideas about life and death, the soul and immortality, which will concern us later.

In particular, a two-kinds-of-stuff view lends itself naturally to a flat rejection of artificial intelligence in its most ambitious sense. Human beings cannot make souls, so the machines human beings can make cannot be given souls. And that is that.

One-kind-of-stuff views

Others have not been tempted to think of life in terms of two different kinds of stuff. The Hebrew Bible, in Genesis, describes a human being as a 'living creature' or, more accurately translated, breathed-into flesh; God's breath gives life to the bodies he has made. There is no 'double stuff'; just one kind of stuff: matter (or energy, or whatever single preferred description we choose).

Modern one-kind-of-stuff views usually treat the mind as an evolved property of the body. Without a working brain, a functioning sensory system, and so forth, we would have no mind (and without language our minds would be much poorer and less powerful than they are).

Histories of Humankind

Our attitude to two-stuff or one-stuff views of the world alters the way we talk about ourselves. Christians, for example, are to be found in both camps, and their beliefs differ accordingly. Some choose to ignore biblical teaching about the resurrection of the body and to think of everlasting life in terms of souls without bodies mixing with one another in some unclear way in a world called heaven. Others take the resurrection of the body more seriously and imagine that we will go on relating in some sense or other much as we do now, but freed from the pain and suffering of our earthly world.

Most of us probably find it easier to believe in an afterlife if we are two-stuff people; the idea of a soul journeying quietly to heaven to begin a new existence while the earthly body is buried or cremated seems to make much more sense than any idea of a new body in a new world. (Although it is silly to argue that just because Great Aunt Mary's body is still in the churchyard, she cannot be resurrected with a body in heaven.

Only a very literal belief treats the resurrection body as if it were made of the same kind of stuff as our earthly bodies. Even if we were to find bones buried in Israel which were demonstrably those of Jesus – although I am far from clear how we could ever be sure – belief in the resurrection would not thereby be rendered absurd. It might be harder to believe in the Empty Tomb, but I at least have never regarded the absence of Jesus' earthly body as necessary for him to possess a resurrection body, any more than that would be true of Great Aunt Mary.)

The resurrection of the self makes very little sense without the resurrection body on the grounds that without such a body the self would have no opportunity to manifest itself to its-self or to others. Without some means to tell stories about ourselves (in words and actions which are typical of those selves), and something to tell stories about, we have no opportunity to be or to develop as selves. Part of what makes me what I am – an important part – is the collection of stories I tell and believe; they reflect the orientation of my brain.

What kinds of stories have we told about ourselves? Plato told a story of imprisoned souls searching for their ultimate release and freedom; the Hebrews thought of life in terms of living bodies for whom the after-life was at best a shadowy affair as we roamed around in Sheol. The Christians did not believe in existence before earthly existence (as Plato had taught), but they certainly believed that the sufferings of their bodies would be the key to everlasting glory in a place of honour in heaven. During the seventeenth and eighteenth centuries, reason took over and we thought of ourselves as thinking things (although often the thinking thing remained an old-fashioned soul). With Darwin we started telling a story which made us much more like the other animals, and from that modern humanity has tended to assume that we have no souls, just efficient brains that produce minds to help us to survive.

The latest in this long and complicated history of stories that we tell about ourselves ('myths' as some would call them) is that we are 'really' just machines running programs. And at first glance it is not at all clear whether such a view sees human beings as made of one-kind-of-stuff or two.

What is clear – although quite why it is clear and whether it is as clear as it seems needs careful thought – is that the suggestion that we are just machines running programs offends most of us greatly. We feel that if this claim is true, we have somehow lost something very important, even vital, of what it is to be human. It is of course possible that such a view is wrong. It may be that we can (or at least should) live happily with the idea that we are machines because nothing at all is lost if we do. But at first sight most of us feel instinctively that it just is not so.

A dual aspect theory

In a dual aspect theory, there are not two kinds of stuff in the universe (material stuff and mental stuff), but there are two distinct sorts of being. There are things which are capable of and deeply prone to look out upon the world (however crudely), things with inside-out orientations as well as outsides (such as organisms), and there are things with just outsides (such as stones). It makes no sense to ask what the inside-out perspective of such things as stones might be. Although some, such as Teilhard de Chardin, have wanted to extend the notion of an inside-out world as far as atoms, in my view such ideas are scarcely credible, even if they are intelligible. Self-awareness, inside-out-ness, consciousness and all the other features of mind are only ours because we enjoy the properties of complex brains. In order to believe that atoms have inside-out orientations, we have to be prepared to believe that mind is not a feature of some organic structure, and therefore we have to be two-kinds-of-stuff people. I do not believe that we are two-kinds-of-stuff things; I do not believe that there are two kinds of stuff.

By the 'inside-out' of an organism I do not refer to what can be seen by opening it up, but to its orientations, which render it fundamentally different from a stone. The inside of an organism is its inside-looking-out-ness, and the perspective of that organism on all other things constitutes an outside-looking-in-ness. There is a directedness about these orientations, something irreducible to mere physical, chemical and biological description. We would have no reason to believe in the existence of such directedness in organisms but for the fact that we experience it in ourselves.

I am a brain looking out; you see my brain looking in (when you dissect it or open up my skull or take an image from a scan). These two ways of looking are completely different. If you describe me as a brain without including any account of my inside-out life, you fail to describe almost everything that is really interesting about me.

Most neurophilosophy (as the philosophy of brains and minds is sometimes called) deals only with aspects of the brain as it can be understood from-outside-looking-in, so do most computer programs which attempt or pretend to simulate human intelligence. Any artefact which is a candidate for treatment as if it were human or conscious must have an inside-out world. And that applies however intelligent it may be. Without this criterion, how intelligent a machine may one day be is on a par with the question how fast an aeroplane may one day fly (although more interesting).

Spinoza, Max Planck, Roger Collingwood, Michael Polanyi, Teilhard de Chardin, Donald Mackay, and Thomas Nagel, among many others, have adopted dual aspects theories.[1] Few, if any, of these writers, however, have attributed sufficient significance to the sense of orientation. A brain's orientation towards the world is a thing unique to that brain. Nothing is lost (and a good deal gained) by identifying this orientation with what we regard as the mind. Accordingly, the mind is the world-orientation of a body. (We might just say 'of a brain', but since the body affects the brain and therefore the mind, it is better to say 'body'.) Of course, this means that without brains there are no minds, but it does not imply that brains are sufficient to guarantee minds. We know enough about non-genetic, environmental influences on brain development through the plasticity of the brain to be able to see that the *evocation* of minds arises from the interaction between bodies and the world. In fact, it is no doubt true to some extent that mild and even severe pain play a significant part in this evocation; we are stimulated and nudged into being by threats and dangers. But the world calls minds into being from the inside rather than manufacturing them from the outside. And that is one of the fundamental differences between a mind and a program: minds are not switched on, programmed, or manufactured; they are grown.

The terms I have introduced (inside-out, outside-in, for example) do not explain the relationship between mind and body, and they do not give a scientifically useful account of how the brain comes to generate that experience which we call mind, but they do contribute to our store of metaphors in a way that prevents us from thinking of the brain simply as an input-output device or of the world of the mind as consisting only of brains which science can describe. In its double aspect, the mind-brain characteristic of higher organisms renders any such approach inappropriate.

The integrated notion of the mind and the body that I have so far insisted upon cuts away Greek dualisms and places us back in a Hebrew way of thinking about humankind in which each human being is a living creature. The human mind is neither inherited from nor coerced into being by the world; it is *called* into being by the world.

Once the inside-out/outside-in distinction is established, many confusions in the discussions of computer science and consciousness become easier to identify. Descriptions of computational power that yield impressive output, for example, need not on this reading be associated with any kind of inside-out world at all; they need only be black boxes with input-output devices attached. The computers involved and their capabilities are not different in kind from fast cars, aeroplanes and rockets; they just represent ways to achieve an end at faster and faster speeds. And even by introducing new concepts like parallel processing, neural nets, genetic algorithms and so forth, we still find ourselves dealing with apparatus that just solves fairly tightly defined problems more efficiently than we could solve them otherwise.

In other words, if we learn to see a computer – however mysteriously and cleverly it may seem to solve puzzles – rather as we see wide-bodied jets (efficient ways of getting a job done), a lot of the mystique surrounding them (and a lot of the anxiety they generate) will evaporate.

Once we have accepted the need for a double story – an inside and an outside story – if we are to give anything like an adequate account of what it is to be human, we need to add a third class of behaviour which any artificial candidate for consciousness would need to satisfy: that which realizes the inside-out perspective of a living thing, which generates an inside-out perspective upon the world from inside the machine and moves a step closer to producing a machine which is genuinely alive.

A problem with consciousness

Consciousness is rather like a bar of soap. Until you reach for it in order to try to grasp it, it seems perfectly easy to handle; but as soon as you try to hold on to it, it slips from your grasp.

In part, this is exactly what we would expect: consciousness is less a something than a process; less something to be focused upon than something we use to focus upon other things. It is elusive because, in trying to look at it, we cease to use it to look at something else, and it passes from view and evaporates.

What we call self-consciousness fares no better. We each have some fairly good idea of who we are, but as soon as we try to stop to grasp the essence of who we are we find that we begin to dissolve. We can only be conscious, or conscious of the self, while engaged in something else. Consciousness is condemned to be the bridesmaid forever, and never the bride; always on hand to help proceedings along, but never the focus of attention itself. As David Hume put it, 'For my part, when I enter most intimately into what I call myself, I always stumble on some particular perception or other. I can never catch *myself* at any time without a perception.'[2] This may explain why for all the attention that has been lavished upon it, and for all the books written about it, consciousness remains elusive. We know what it is to be conscious, but not in what that being conscious consists.

As we shall see in Chapter 5, it is nevertheless a permanent preoccupation with most of us to be able to grasp which object in the world each of us is. 'Who am I?' is a question we all ask, and which we seem to ask with increasing frequency and intensity as we grow older. The death of a friend, a parent or, most tragically, a child, brings us face to face with the limitations of our own existence. One day we shall each cease to be, and that means 'cease to be conscious of being alive' as well as being physically dead.

For most of us, the essence of life centres upon being mentally aware, and a principal objective of most religions is to explain how that awareness might conceivably survive our physical death. But it is not at all clear in a world dominated by scientific imagery and doubt, that such a possibility remains believable. Do computers afford us a way to overcome the difficulty? Are we really like programs running in machines? What kind of program is it that could be conscious of itself?

What is consciousness?

The simplest way to think of the nature of consciousness is to imagine it in terms of the brain's ability to be present to itself. The five senses (sight, hearing, touch, taste and smell) each register in the brain. But add the brain's ability to be present to itself by a feedback process (awareness through memory, newly thought words and so forth) and the mystery of the mechanism of consciousness largely disappears. See Figure 5.

What also disappears is any strict boundary between human and other animal consciousness. Animals register their five

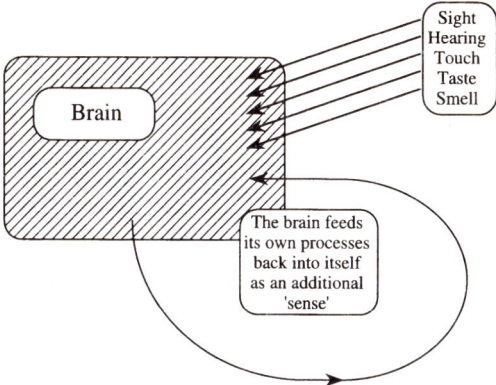

Figure 5 The brain monitors itself.

senses, and no doubt their brains are present to themselves through memories as well (to a greater or lesser extent).

Of course this does not explain consciousness, if by 'explain' we mean 'give us a way by which to understand how it is possible', for it simply attaches the brain's feedback loop to its normal sensory abilities. That is, it says that because the brain registers the products of the senses (interpreting them as it does so), there is nothing intrinsically more mysterious about the fact that the brain registers its own presence than is mysterious about the fact that we register visual images, sounds and the like. Of course, that is very mysterious indeed (and I say 'mysterious' with some care).

This simple feedback loop enables us to explain only to the extent that it helps us to see why, in practice, when we set out on a repeated loop (thinking about thinking about thinking ...), we are dealing with an illusion. Each cycle depends upon our ability to present to our current conscious selves those things we produced earlier (whether 'earlier' amounts to 'moments ago' or 'years ago'). The actual process remains a mystery; the structure does not.

An explanation of consciousness would involve a description of what it is about the brain that enables it to register the products of the senses as qualities (*qualia*). Having attained that, we would then have no further difficulty with consciousness. But without the dual aspect theory of mind and body I have outlined in this book, a theory with a long and distinguished history, we cannot come anywhere near an adequate account of our awareness. (Compare 'The *qualia* problem' in Chapter 1.)

One of the most important properties of feedback is called 'recursion'. Recursion involves the way a system's next state depends upon its previous state or states. This may not seem very remarkable or interesting, but it gives rise to one of the simplest kinds of unpredictability. The brain is a recursive system because its next state depends upon countless previous states (since it is a parallel-processing organ, each of whose parts may influence states far distant into the future).

The states of the brain are successively affected by its own past states (including states produced by those processes which were started long ago – as for example when we suddenly find a solution to a problem or a name 'popping into our heads' long after we stopped trying to solve or remember them). And it also shows another aspect of the 'correlation' that must occur if the brain is to assemble all the components in the solution to a complex problem at the same time. (It would be no good, for example, if in trying to carry a fork full of food to our mouths, we solved the problem of how to open the mouth much more slowly than the problem of how to raise the arm. We take our mastery of such synchronization for granted, but the brain solves this and far more sophisticated coordination problems every second of the day.)

The feedback process also helps to throw a little light on the experience of blankness when we cannot remember a name or solve a puzzle. Those 'I don't know how to go on!' moments, where we have exhausted all the subterfuges we usually employ to coax information out of our fallible memories, correspond to states where the brain, however much information it is feeding us along our looped internal sensors, is not giving us the information we are 'looking for'. The experience is just like waiting for someone at a station, searching for a familiar face in a crowd. Vast amounts of information are being presented to us; none of it triggers a resonance sufficient to mark recognition of a familiar face. 'There he is!'

It has long been thought that the solution to the mystery of consciousness is to be found somewhere in the realm of feedback loops. Perhaps the most famous book on the subject is Douglas Hofstadter's *Gödel, Escher, Bach* (Harvester, 1979). The whole book is a celebration of the strange loops that we encounter in logic, art and music as we play with our ability to reflect upon ourselves, to re-enter the same processes again and again. But Hofstadter, like other philosophers of mind and computer scientists before and since, tries to explain the mind/brain relationship solely in terms of the outside world. This is to fail to recognize the profound and crucial difference between observing a brain as a biophysical organ, and being a brain. Yet if Hofstadter and his readers were not 'brains', the whole communicative enterprise upon which he sets out would be impossible. (This is another way of saying that if there were only outside-in processes in the world, there would be no thoughts, no meanings, no sensations of colour, hearing, touch, and therefore no *qualia* and no consciousness.)

It is also another way of saying something which perhaps sums up the message of this book more succinctly than anything else. Since the only way we know that brains possess what we call consciousness is that we *are* brains, and therefore know brains are conscious, it may be that we shall never be able to tell whether a computer is conscious without being able to *become* a computer which, precisely by virtue of being that object which is conscious, would know that computers can be conscious.

We know brains produce minds because we are conscious brains with minds; we could only know whether computers could produce minds if we were a conscious computer with a mind.

What seems certain is that a conscious computer would need to be equipped with a sophisticated feedback capability, an advanced capacity for coordinating disparate processes, and a set of experiences acquired through sensory organs that would entail a genuine degree of emergence, a genuine education, a genuine growth.

We are brought back to a fundamental philosophical problem, which has been touched upon elsewhere: how – other things being equal – could we hope to tell whether a computer was conscious and possessed of a mind other than by taking its word for it in conjunction with interacting with it in the kinds of ways that persuade us that our fellow human beings have minds like our own? (And how would a computer with a genuine inside-out world come to acknowledge that human beings also possess such an orientation?)

Of course, some creatures which are biologically human behave in ways which lead us to doubt whether their minds are fully formed. Human sin, such as in mindless violence, cruelty, destructiveness, rape and murder, may give us reason to ask whether in fact the distinction between the biological and the mechanical is as absolute as we might wish to suppose. Could it be that computers will one day be better inhabitants of this planet than we have been ourselves? (Many of those involved in artificial intelligence exhibit a profound pessimism about humanity and its capacity to redeem itself. A Christian, of course, might well share that pessimism insofar as we hope to redeem ourselves.) And in that case, does human mind and consciousness only represent the latest (rather than the final) stage in an evolutionary process through which God plans to bring into existence the perfect creature? There are certainly plenty of suggestions in both science fiction and the statements of the more optimistic (or should it be pessimistic?) artificial intelligence gurus that robots and artificial life will be better, more responsible citizens than human beings have been.

The problem of creativity

Perhaps the most deeply felt and popular objection to the idea of artificial intelligence, and particularly to the idea of an artificial person with an inside-out orientation – a mind – is that such a machine could not be creative.

To some extent Maureen Caudill has already given us one reason to suspect a fallacy in this argument: if we regard human beings as creative, and if we accept that brains are neural nets, then artificial neural nets can be creative too.

It helps to consider in a little more detail how creativity arises as a feature of the human inside-out world.

From the diagram we used for consciousness before, Figure 5, in which we saw that the brain's ability to feed back into itself and be present to itself is the fundamental process that permits consciousness, we can see the way in which creativity presents itself to us, even if we cannot explain its origins as simply. See Figure 6.

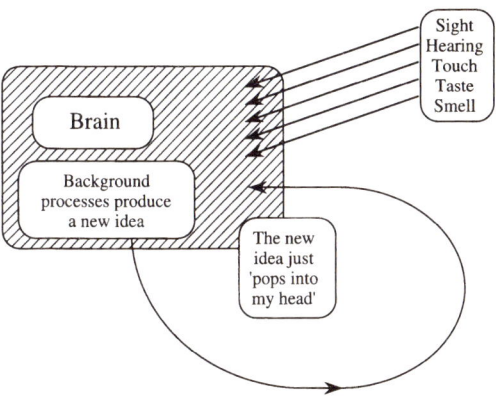

Figure 6 An idea arrives from the background processing in the brain.

Figure 6 shows how the brain, by being able to be present to itself, adds to the normal five senses (sight, hearing, touch, taste and smell) a sixth kind of input. This input involves memories which mimic new sensory input (as in vivid dreams and poignant recollections – smell is often regarded as the most powerful sensory memory in this respect). But it also involves new ideas which do not originate entirely in past experiences, being created afresh by processes below consciousness before they 'pop into our heads'.

The 'new idea' is nevertheless not merely a matter of feedback; it is also experienced, and as such joins the multitude of other *qualia* competing for my attention. It would be perfectly feasible to generate feedback in computer circuits, but there is no guarantee that the computer would then experience the products of that feedback.

What is more, if the new idea is sufficiently arresting to gain my attention (and how, precisely, does it do that?), I will begin to work on it by trying to increase its sense of rightness. And in doing that I will make use of feelings of resonance that draw upon the vast extended field of meanings available to me through my internal and external senses in the frame of reference that is my experience of the world and the 'kind of person that I am'.

Figure 7 shows how an idea which first 'grabs our attention' may not be quite right, and as a result is reprocessed. For example, in writing a book an author may be trying to find a way to spell out a particularly important and difficult idea.

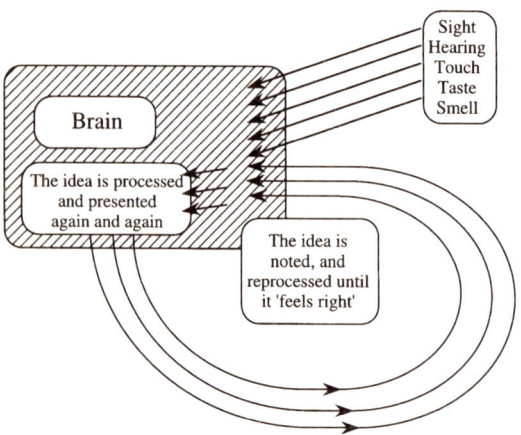

Figure 7 The idea is refined until it 'feels right'.

Many sentences occur to her, but she rejects them until one arises which 'feels right'. Cycling through streams of such sentences until the right one appears (and sometimes it never appears) involves trusting that the background processes in the brain will work. When the sentence we need in order to express an idea just will not 'come out right' there is very little that we can do actively to remedy the situation. We must just be patient.

When I am satisfied that the feedback loop has produced something good enough, I may give expression to it in a word, a sentence, a book, a piece of art, a design, a dress, a quilt, or through one of the countless other media that human beings use to communicate. As something refined and developed inside me, the resulting creation will be very much a part of my person, virtually an extension of my self.

Reading and writing

Writing is an example of creative, inside-out activity that a computer might engage in. I want to explore a particular relationship between writing and reading by supposing that a computer which has been programmed to write poetry produces a poem which I find full of meaning. Since such poems are usually taken to be signs that their authors have some deep insight into the nature of existence, would I be right to infer from the computer's poem that it has an inside-out world, that it has a profound insight into the conditions of existence, and that it should be treated with as much respect as a human poet? Or should I reserve that accolade for its programmers (even if they proved incapable of writing a novel or a poem of equal quality on their own)?

It is too easy to dismiss this 'thought experiment' by saying that a computer could do no such thing, and to do so would be to miss my point. I am not in this instance concerned about whether such a thing is possible, but with what it tells us about reading and writing and our attitudes to both. (I shall use poetry as my example, but the same can be said of plays and novels.)

It is easy to assume (and many of us do assume) that if we can see a meaning in something, its author must have intended that meaning to be there. Children, when they first encounter literary criticism and serious verse, will frequently be heard to protest (after some particularly imaginative interpretation by a teacher) that they are sure the author 'couldn't possibly have meant all that' when he wrote the poem.

And of course they may well be right if by 'meant' they mean 'consciously and deliberately and premeditatedly intended' the poem to have that meaning. But to suppose that such deliberate intentions are required in order for the author to be credited with having intended such meanings is almost completely to misunderstand the creative process. On the contrary, we value the capacity of writers to put things into words in ways which somehow point towards (even if they do not and cannot encapsulate) their seemingly infinitely rich seams of meaning. Great poems and novels always mean far more than their authors could ever have brought consciously into their own minds. That is what makes them great. (Remember in this context the point made earlier about artificial intelligence: we should not be too ready to credit it all to the programmer.)

What this means in the present context is that a computer which produces a poem that we find full of meaning must, if the computer is to be credited with having dimly perceived those meanings (rather than its programmer), have a capacity for appreciating the inherently ill-formed and vague meanings which exist at the periphery of poetic language. Failing that, it need have done no more than respond to a program which is to the programmer rather as a novel to its author; the program embodies rich insights into the nature of poetic language, all of whose ramifications and possibilities the programmer could never consciously have envisaged. What is more, the program produces output which is enhanced by the imaginative endeavours and abilities of its readers. The computer itself need have no understanding whatever of what it is doing (just like the man whom we met in Searle's Chinese Room who manipulates Chinese ideograms according to sets of rules without understanding what those ideograms mean).

This being so, it throws the burden of decision about whether in fact a machine has an inside-out awareness of itself into the sharpest relief, for in any given instance it may be impossible to disentangle the output (the behaviour) of the computer from the creative input of its designers and the imaginative capacities of its human assessors. For example, Rainer Maria Rilke's poetry evokes echoes in us because we can regenerate caverns of possibility from our own experience within which those words can resonate. Remembering my own parents' deaths relatively recently, I shudder with a mixture of thankfulness and awe whenever I read the wonderful line with which he commences his 'Requiem': 'I have my dead, and I have let them go.'

How much of the power of those words is Rilke's, and how much mine? And what would it require for a computer to experience, from-inside-out, that involuntary shudder? It would, I think, require the computer to have developed something akin to the human soul, that realm of the self that lies always on the periphery of consciousness, ready to surprise us.

Summary

The fundamental distinction that has been made in this chapter is that between an object which can be accounted for entirely in outside-in terms, and an object which requires both outside-in and inside-out descriptions.

Technological achievements in artificial intelligence which remain in the first category present us with no new issues, even if they present new challenges. It is only when we begin to countenance the idea that a computer or an android might require both kinds of description that something fundamentally new will have been brought into the world. At the time of writing, and for the foreseeable future, no such developments seem remotely likely.

3 – PROBLEMS WITH WORDS

One of the problems we face with a question like 'Is the brain a machine?' is that 'machine' is used in so many different ways. Yet the word alone carries enough bad associations for us to shrink instinctively from the idea that brains are machines (unless we are paid-up members of some artificial intelligence club). This vagueness has been exploited by some writers who use a particularly naughty ploy: they agree to use a word in a precise but rather unusual sense while we are on the alert, but subtly shift the meaning later to embrace everyday senses of the word not covered by their initial definition. This move is so common in the literature on artificial intelligence that it needs a succinct name by which we can refer to it wherever it rears its head. I shall call it the Sliding Definition Ploy (SDP for short).

For example, Frank Tipler, in his stimulating but bizarre *The Physics of Immortality*,[1] performs a wonderful example of SDP by defining life in what he styles 'physics language' in terms of entities which code information. 'Thus "life" is a form of information processing, and the human mind – and the human soul – is a very complex computer program' (124). But this is to make life self-evidently a computer program, for computer programs are coded information. Nothing much can then be deduced about whether life is a computer program, since we have started with that as our definition. Again, Richard Dawkins, author of *The Selfish Gene*,[2] employs at times a definition of life which permits him to say that a motor car is alive. This should warn us that some kind of SDP is to be perpetrated, probably something along the lines of 'Well, if even a motor car is alive, then why not a computer?' We should simply refuse to grant the initial premise: to use the word 'alive' to describe a motor car is to stretch the word's meaning to the point where it becomes useless.

The dangers of falling victim to some kind of SDP are great where words name inherently vague notions in the first place. A 'machine' seems to be a clear concept until we start to ask about its boundaries. A motor car is certainly a machine; so is a typewriter; so is the product that washes the clothes. A 'machine' seems to have the quality that certain inputs produce certain clearly definable outputs as a result of the pursuit of clearly defined rules of operation (operational principles). A machine also usually takes a relatively small or simple input and makes it larger in some way by allowing us by, for example, throwing a switch, to perform a task that would otherwise involve considerable labour. (It is this quality that allows us to describe a lever as a very simple machine.)

Even if we allow this as a starting definition, we soon run into trouble. We tend to call computers 'machines', but are they machines in the sense stated? A given input certainly always produces a given output, but are we always in a position to say what that output will be, and if we are not, have we widened the definition of 'machine' to the point where it starts to lose its precision? Have we even started to perform some version of SDP ourselves?

This is not a book about philosophy of language, so the problems associated with vague definitions need not concern us directly. What does need to claim our attention is the motive for asking a question such as 'Is the brain a machine?' at all. The real point of that question is, I think, to ask the question, 'Is there anything special about brains?' (which betrays the assumption that there cannot be anything special about machines).

Of course, this is itself a very vague question. We all want human beings (and, in these enlightened days, other higher forms of life) to be special in some way, and (human) brains seem the most likely candidates to make them so. But we lack a persuasive way of explaining the action of brains that does not rely upon machine-like operations. We use terms and concepts such as cause and effect, and so we lack a way of explaining brains that does not rely upon machine-talk at all. Yet there is no reason in principle why processes in organisms, and especially brain processes, should actually fit into language based upon machine concepts. It is just that in the absence of other well-worked-through concepts we tend to fall back upon familiar metaphors and other intellectual tools. Those in the 'strong AI' camp have to some extent managed to persuade some of us that our brains are like digital computers for want of a competing set of concepts that we might use.

One way to flesh out the question, 'Are brains machines?' is to ask whether they can be understood completely (and therefore adequately) in terms of rules of procedure and other operational principles borrowed from mechanical models. If so, there is nothing special about brains beyond their size and complexity, and one day we shall find a way to build a computer that replicates all their functions.

Unfortunately, this 'beyond their size and complexity' hides a multitude of further problems. The assumption that by increasing size and complexity we introduce nothing essentially new that, so to speak, cannot be dealt with just by using larger numbers, has been found to be false. Size and complexity can involve sudden, unexpected and precipitate changes in the nature of an object which invest larger and more complex things with properties that were unthinkable in their smaller siblings.[3]

With these preliminary thoughts before our minds, we can now proceed to try to pin down what we mean by 'machine', 'life' and 'intelligence'.

Machine

A machine is an object which has at least the following two properties.

60 · *God and the Mind Machine*

- *It confers an advantage on its user.*

  In other words, machines enable users to do something they could not otherwise do. A lever is a simple example: by applying a small force but moving it a large distance, I convert it into a large force that moves something over a small distance.

- *It embodies a close, reliable and traceable connection between the input and the output.*

  In other words, when a machine does something in response to some input stimulus (such as the pressing of a button), it is always possible to trace the output of the machine back to that input stimulus, and that same input stimulus must, other things being equal, produce exactly the same output. Such-and-such a force applied to a particular lever will always produce such-and-such an output force.

  Some would wish to include this further condition.

- [A machine always commences its operation from the same point and cycles through the same (possibly very complex) set of procedures until it is switched off.]

  The square brackets indicate that I do not regard this as an essential property, for a reason that will almost immediately become clear.

A 'digital computer' is a machine because it satisfies each of these requirements (including the last). For example, it calculates more quickly than a user, and given input always produces the same output, other things being equal.

The question whether a human body is also a machine is actually unanswerable, although it seems likely that it is, because the 'other things beings equal' criterion (which is absolutely essential, since if it fails neither a simple lever not a digital computer may always produce the same output for the same input) cannot be satisfied in any experimental arrangement. In other words, it is simply impossible to set up any human body in two different experiments in such a way that all things other than whatever stimulus we might supply are equal. (This remains true for much less complicated bodies as well.) No human body could possibly satisfy the third requirement.

However, it seems churlish to deny that *if* this 'other things being equal' condition could be met, human bodies would qualify as machines. But it would be equally churlish to deny the significance of the fact that such a condition cannot be met. That significance must include the inherent unpredictability of human behaviour (since a condition necessary for predicting that behaviour – that all other things are equal – cannot be achieved). It *may* also include the observation that existence in a temporal world in a state of permanent flux guarantees that unpredictability. Digital computers do not in general inhabit such a 'world'.

It may seem startling that we seem obliged by this discussion to admit that *if* a human body were to be placed in an absolutely identical situation in exactly the same state as it was before, it would necessarily behave in the same way. But this is actually completely trivial: the word 'exactly' must bind us exactly, and so there could be no conceivable reason why we should behave differently.

Life

This discussion of the nature of a machine leads immediately to a corresponding (but more elaborate) notion of life.

- *It sustains itself by processing energy from its environment.*
- *It reproduces.*
- *It evolves.*
- *It is born and it dies.*
- *It copes with a constantly changing and possibly non-repeating environment.*
- [It exhibits agency (self-direction).]

The Sliding Definition Ploy works, when it works, largely by omitting one or more of these characteristics. Frank

Tipler's definition of life in terms of information processing could easily replace the first, but it does not necessarily satisfy the second, third and fourth. And a motor car does not satisfy the second, so Dawkins' argument that a motor car is alive fails.

A digital computer is not alive because it satisfies the first only if we interpret it generously.

Are products of digital computers alive? Is it reasonable to describe the things we find inside digital computers (such as so-called 'viruses') as 'artificial life'? Artificial 'life' seems to satisfy the first four, but it may or may not satisfy the fifth and it certainly does not satisfy the last.

Are conditions five and six too strong? Suppose we apply them to something like a plant, which we consider to be alive in normal parlance. Here, the fifth does seem adequate: plants do respond to constantly changing conditions which may never repeat exactly. But is it realistic to describe them as exhibiting agency? A better (because less demanding) requirement than agency would be that life is governed by a mixture of internal influences (such as genes, needs, memories) and external influences (such as threats, food supplies, temperature). So let us replace the sixth requirement with:

- *Is governed by both internal and external influences.*

This embraces agency, but does not restrict life to those organisms which are capable of it.

Much of the 'artificial life' that is discussed in the computer literature (including 'The Game of Life' invented by John Conway) fails to satisfy the last condition even in this weakened form. Because it is generated and controlled entirely by externally imposed rules, and has no internal structure of its own, there is nothing 'inside' each 'organism' that might direct it.

The original version of the last criterion belongs elsewhere, in the definition of intelligence.

Intelligence

We shall describe an object as intelligent if it exhibits the following properties:

- *It responds appropriately to new situations.*
- *It solves problems in a certain (perhaps highly restricted) set of circumstances.*
- *It exhibits agency (self-direction).*

Once again, some would wish to add another criterion.

- [*It is aware of what it is doing.*]

The first condition reflects what might be called the 'latency' of intelligence: that it, so to speak, 'waits' to reveal itself; it is ready to be called upon when needed. To some extent this latency may seem to conflict with the second, but 'newness' and 'a restricted range of circumstances' need to be taken together. An effective chess-playing program, for example, responds appropriately to new positions on the board, but it does not solve problems outside the chess realm. Human intelligence does not generally manifest itself in any and every set of circumstances to an equal extent. (I am better at solving mathematics problems than translating Mandarin Chinese, for example. Someone else's intelligence might show itself the opposite way round.) This is reflected in the second requirement; an object's intelligence is measured by the way it solves or fails to solve problems in quite specific areas.

A notion of intelligence that would continue to hold in the absence of all such abilities strikes me as empty. It would involve saying something like 'This woman is very intelligent, but she cannot demonstrate her intelligence by solving any problems at all – she doesn't have that kind of intelligence.' I do not find this a very helpful notion. The second condition reflects the need for intelligence to be earthed in concrete situations.

The third and fourth criteria point towards just the kind of inside-out, outside-in distinction upon which I have based much of this book. An 'agent' behaves in ways which are not predictable. It expresses its preferences through its responses to new situations. An intelligent computer would similarly respond appropriately to new situations, and solve problems, but it would also have to satisfy the machine condition that its behaviour be traceable. That is not a condition that life can in general satisfy. This refusal to expect complete explanations adds to life's mystique. The behaviour of an agent is not in this sense 'automatic' or fully explicable, even if only because the 'other things being equal' condition is seldom, if ever, met. In other words, because we cannot specify exactly what the initial conditions were for any organism – where it 'starts from' – we cannot hope to explain its behaviour completely.

'Why did you drink tea rather than coffee this morning?'

'I just felt like some tea.'

This is not an explanation, but what would be?

An agent, it seems, must exist in an environment in which neither an organism nor any observer of that organism could possibly give a complete specification of all the conditions that bear upon the way it behaves. This is another way of saying that whether or not organisms are machines is unanswerable in principle. And whether or not an artefact is intelligent may in the end boil down to a non-trivial application of the familiar principle, 'It takes one to know one.' Intelligence may be more something we recognize and exhibit than something we have.

Of course, there will always be sceptics who will deny any computer intelligence even on these terms. They will prefer to argue that whatever intelligence a computer program exhibits is really 'just' the intelligence of its programmers. I have some sympathy with this view, but it overlooks two important interlinked considerations.

It overlooks the phenomenon of complexity, that systems based upon apparently simple rules can generate levels of complexity that were not anticipated by their creators.

The argument also begs all the interesting questions by arbitrarily cutting the chain of responsibility at the programmer. To credit a programmer with all the demonstrations of intelligence that arose from his program is only as justifiable as, say, crediting my teachers – or, more reasonably, the whole universe (for where otherwise does the list end?) – with the intelligence (such as it is) that I am using to write this book.

And this points, of course, towards the creative activity of God. To deny the products of intelligence their own intelligence would be to deny ourselves intelligence. But does it make any sense (or, more to the point, is it a remotely helpful way to use language?) to say that all the intelligence in the universe is simply the outworking of God's initial creative intelligence? There is a profound and important sense in which that is true, and I do not for a moment deny it. But our intelligence is more a matter of taking advantage of a possibility God's creative intelligence made available. To deny this would be to make God directly responsible for everything human beings do, with all the attendant problems that would create.

ALGORITHM

The word 'algorithm' means a sequence of rules which can be followed to complete a task. Algorithms are detailed sets of instructions in a language which a computer (or a human being) can work through without needing to understand anything that is not supplied in those instructions beyond what the terms used mean. For most practical purposes in what follows, the word 'algorithm' can always be replaced with 'procedure' or 'program', despite the fact that there are technical distinctions between the three.

Simulation and emulation

Those working in the artificial intelligence field recognize the difference between a program which employs techniques completely different from those employed by a human to appear intelligent (including, say, setting human intuition over against algorithms and computational power). Such programs 'simulate' intelligence. Programs which as far as possible attempt to reproduce intelligence by employing techniques that mirror those used by their human models, can be said to 'emulate' intelligence. (If I were to build a copy of the Taj Mahal out of cardboard, I would be simulating it; if I were to build it out of the same materials as the original, I would be emulating it. Simulation, in other words, attempts to mimic; emulation attempts to reproduce. It is unfortunate that the AI literature tends to use 'simulate' indiscriminately.)

A computer which sets out to do no more than equal or better human performance at some task could be set up in a way which attempted either to emulate or to simulate the performance of the brain. An emulation would have hardware requirements: the computer would need to be built in a way which reproduced brain structure, not just brain function. But an approach which was content merely to appear equal to or better than the human brain could employ any technique at all.

Simulation and emulation are both qualities which are and need only be observed and assessed from-outside-looking-in (compare Chapter 2) by human operators sitting at computer terminals or observing robots and measuring how well they manage a task. Both, although they mark an important distinction, are features of the machine which do not require the generation of an inside-out world. (Nobody cares whether the chess-playing computer is enjoying the game; the objective is to win the competition for the team which made it.)

How far simulation and emulation will manage to travel into realms hitherto thought the exclusive domain of (at least) organic life and (at most) human creativity, is a fascinating and alluring question, but it is not different in kind from other purely technological questions about things such as how fast we may one day manage to move (how close to the speed of light?), how close to absolute zero we may one day reach, how comprehensively we may one day model the universe, or how successfully we may one day generate new life forms by genetic engineering.

Of course, as each frontier is pushed back, those concerned to identify the uniqueness and importance of humanity with the very skills that science has eroded will find themselves 'on the back foot'. But we need not bat on that wicket at all. I do not think we should make extravagant claims about other areas of human creativity that have similarly been identified as peculiarly human (the composition of music, the writing of novels and poetry, and the production of paintings and sculptures). Inept as most of the attempts at novel-writing and poetry by machine have so far been, we have seen often enough already that only a brave person laughs at these early setbacks. Time is on the side of the simulators and emulators. (There seems little doubt that some of the formulaic novels that are currently written will one day be written just as well by machines.) Whether (and to what extent) we would be prepared to read (still less enjoy) a novel or a poem written by a computer if we knew that it was written by a computer, depends upon a clearer understanding of what it is we are doing when we are being creative in (say) writing words, or being creative in reading them. (See the end of Chapter 2.)

Explanation and Prediction

In this section I shall outline a distinction between two kinds of story. This distinction is important in considering computers (and it is also important in considering such things as evolution, although I shall do no more than point to its importance there in passing). Essentially the distinction involves recognizing that predicting the future and explaining the past are not mildly, but fundamentally, different kinds of story-telling.

Telling stories about the future involves throwing forwards or 'projecting' our present knowledge and trying to see how things might turn out. It involves trying to estimate which of many things will happen on the basis of our best understanding of ourselves and the world. In the case of a digital computer, we know that its future will always involve following some kind of program with strict rules (even if those rules sometimes amount to tossing a coin in some electronic way, and even if they also modify themselves).

Explaining events in the past involves throwing backwards or 'retrojecting' present states of affairs to see where they came from. Whereas with projection we may be completely wrong and the set of circumstances we predict may never arise, with retrojection they are already here and we are seeking to understand why they came about.

Projection and retrojection differ because the states of affairs they aspire to are asymmetric. It is true to say that any computer arrives at any state by following a route through its program which can, in principle, always be traced back to the moment when it was switched on. Retrojection is always possible (in principle). But it is false to argue that because a computer always just follows its program, we can therefore always predict its future states, and therefore always project its current state forwards.

Confusion in evolutionary biology is rife in this area. Our ability to retroject each evolutionary development to some previous set of circumstances tempts us to infer that the evolutionary development was inevitable or random. But we just cannot infer that. And therefore, the claim by evolutionary biologists that evolutionary theory explains how everything has come about is logically erroneous. And as soon as that claim is dropped, it ceases to be possible for any evolutionary argument based upon it to deny at least the potential involvement of a divine being in the process.

In much the same way, although we imagine that the course of a program running in a computer can always be predicted, this is actually to confuse projection with retrojection. It is one thing to say that whatever a computer does can be traced back to its program (along a possibly very tortuous track indeed); it is another to say that we could trace the program forwards in the same way. These are the problems of initial conditions and real-time realization; unless we know precisely the state in which the (complex) system begins, and can predict the outcome of every potentially random fluctuation it takes into account faster than the process itself, tracing forward is not possible. Therefore, if the initial conditions are vague and/or the path the program follows is tortuous, or our simulation too slow, the only way to arrive at any distant future state of the program may be to run the program and wait. This is called 'solving the problem in real time'.

On such a supposition, even if we were programs running in machines, it might still be impossible (given even slightly vague initial conditions) to state exactly how we might be in any given 'now' (and correspondingly impossible to specify the states of thousands of millions of semi-independent organismic 'units' at any future time). The specification of the exact state of a sufficiently complex computer system at some future time might, on this reading, be beyond even God.

This can be summarized in the following paradox: the power of the computer arises from and lies in the fact that we constrain its operation so greatly that it always does exactly as we program it to do (we force it to behave predictably in most circumstances); but a computer which hoped to approach a state equivalent to the human mind would have to have those constraints removed to such an extent that it would virtually cease to be a computer at all.

This may seem too neat. Would it not be possible to conceive of a computer which reasoned more logically, argued more cogently, calculated more accurately, and made decisions more precisely and wisely, than a human being? And would not such a computer be superior to a human being? How does the argument I have just given affect that thought?

Such a computer would not be superior to a human being because it would not be making decisions appropriate to the human situation. It would lack all the emotional involvement in the human situation upon which the real importance of our decisions turns. The abstract idea of 'the right decision' only makes sense when we add 'for me' and 'in my life-situation'. A supposedly perfect rational decision made by a computer on the basis of a general appreciation of the state of a human being in the world would not, in fact, be a genuinely rational decision at all. It would not and could not have taken into account the only consideration that actually counts for anything where the rightness of the decision is assessed: that it be mine (and appropriate to my unique situation). In other words, what matters about a decision is not so much that it is right in some abstract frame of reference, but that it is a decision I am happy with. To make such decisions for me, my computer would need to feel what I feel, know what I know, and experience the world as I experience it. It would need, in effect, to be me.

And our notion of 'humanity' in the abstract (free, for example, from all pain and sorrow and suffering and death) may actually be at the root of our readiness to be overwhelmed by the computer metaphor and its ideal of impersonal, detached analytic reason. It scarcely even makes any sense to speak of intelligence in abstraction; intelligence is always located in a context and reveals itself in relation to a particular problem or need. So it may be that the notion of computer reason, by lacking the context in which that reason could be said to have been applied to a particular area of need (and the computer itself has no needs), may not be reason at all, but an empty abstraction as meaningless as 'intelligence' in the absence of anything to be intelligent about or to reveal intelligence in.

That both reason and intelligence seem to evaporate when taken out of real contexts in which they can be seen to be operating makes them similar to an equally important feature of consciousness: that it is altogether clear what we mean by it unless we make the mistake of trying to grasp the notion too firmly.

Artificial Intelligence (AI)

STRONG AND WEAK AI

Although I shall not attempt to take up a position on the spectrum leading from weak to strong AI, it will be necessary occasionally to refer to the extremes.

Strong AI, although defined variously in the literature, boils down to a belief that sooner or later all the characteristics of organic life will be realized in machines of one sort or another.

Weak AI, although it too is variously defined, amounts to the more modest belief that significant portions of those areas of human intelligence which can be given concrete expression will come to be built into machines. For reasons which I discuss in many different contexts, I regard weak AI as almost trivially true (granted a somewhat broad definition of intelligence).

What those in the 'strong AI' camp have not acknowledged and should not acknowledge, despite all the setbacks and disappointments which AI has suffered to dampen its early optimism, is that our inability to build a brain precludes us from building something as good as a brain. And therefore they are profoundly unimpressed by arguments that seek again and again to show that digital computers are not like brains (and vice versa). Their view is, 'Never mind how we do it as long as we do it.' And this really brings us close to the centre of the real issue: is a computer which simulates human intelligence and conversation in a way that is indistinguishable from (other than in being superior to) a human being therefore the equivalent of (or better than) of a human being? 'It's not an important question because it will never be done' is a common, but to my mind not very helpful or constructive, response.

My strategy is to concede that computers may one day achieve unimaginable feats of apparent intelligence and even be able to engage in conversations that are as entertaining as, or more entertaining than, conversations with other humans, and yet to deny that such computers count as the cybernetic equals of human beings.

NEURAL NETS

There are two distinct camps in the artificial intelligence community: those who think that digital computers processing elaborate sets of instructions coded into programs offer the best hope of success (this is sometimes called 'Symbol Processing'); and those who think complexity theory, neural nets and genetic algorithms offer the best chance of success.

A neural net approach, also known as 'connectionism' and 'Parallel Distributed Processing' (PDP) attempts to emulate brain function by building circuits and initiating learning processes which model the ways neurones are linked and operate. Attachments to neurones are given weightings that determine when they fire, rather as occurs in the brain. Neural nets are particularly effective in tasks such as pattern and face-recognition which typically defeat symbol-processing approaches.

One important feature of a neural net is that it can learn to do things through training which configures it in ways that cannot be specified in advance. We can reproduce a given neural net once we have one that works, but we cannot know (before it is trained) what its weightings will be.

GENETIC ALGORITHMS

Another approach is to treat strings of code like the chromosomes in a cell which divide and join up to form new combinations of code. Unlike neural nets, genetic algorithms require us to have some idea of the kind of solution we are looking for if they are to work (otherwise their evolution has nothing to fit). Sophisticated models introduce predators into the environment in which the genetic algorithm evolves to encourage richer variation and so a more thorough search of the space of possibilities in which a solution is believed to lie.

The kind of target we would use in, say, a timetabling algorithm, would be that every teacher had a class, no teacher had two classes, and every class had one and only one teacher at any given time (etc.). The algorithm would then see if it could fit all the teachers and classes together by a process of evolving trial and error.

EXPERT SYSTEMS

One other area of research in AI once thought to offer real hope of breaking new ground is called 'expert systems'. 'Expert systems' is the name of a computer programming strategy, rather than any particular way in which that strategy is implemented. An expert system attempts to store knowledge gleaned from human practitioners in a machine form. One early area of interest concerned doctors' medical expertise, where possible diagnoses were cross-referenced from large databases on the basis of symptoms reported by patients by a program called 'Mycin'.

Two discoveries were made early on in this research: that human beings find it hard to tell what they know; and that it is better to work with one expert than with many. The first problem is very deep; it connects with the Frame Problem we have already met and discussed. A doctor, for example, is not only drawing upon consciously expressed knowledge when making a diagnosis; she reads body language, takes account of vocal inflection, and references possibly long acquaintance with patients and their families. She makes a balanced judgement without consciously processing all this information; her experience sets the frame of the problem. It is, accordingly, hard for her to say (explicitly) what she knows.

Expert systems are also ideally required to give explanations for their decisions. For example, if the medical program suggests that a patient has gallstones after being presented with several symptoms, the doctor would reasonably then ask why the computer had come up with that diagnosis. If the computer were incapable of giving an account, the doctor would feel uneasy; but if it gave a string of reasons with which the doctor could agree, then the doctor might proceed with much more confidence.

A pragmatic question

Religious people tend to want answers to theoretical questions which say 'yes' something is possible or 'no' it is not. The question of the consciousness of an artificial human being, whether we call it robot or android, seems to require an answer: is it possible? Yes or no?

Engineers and social scientists tend to be much more practical. In this instance I think that we can learn something from them. I am pressing the case for an open-ended attitude to the question, 'How intelligent will computers one day become?' I argue that this is a purely technical problem of very little significance for either theology or philosophy. In essence, once we have analysed a problem sufficiently carefully to be able to think of writing a program to solve it, we have probably answered the question of intelligence. It is scarcely very important whether the intelligence lies in the computer or in those who program it. What matters to consumers as members of the general public is how that computer performs and what contribution it may make to their lives.

So let us consider the following practical example of how we might respond to a robot; an example which I believe to lie well within the bounds of realizable computer programming, even if it cannot yet be done. It also illustrates how we tend in the end to solve philosophical questions pragmatically, rather than theoretically.

Let us call our domestic android 'Mavis'. Mavis does not do the housework (doing the housework is a massively difficult task because of all the problems that must be solved to negotiate and manipulate randomly situated and irregularly shaped objects), but she does talk and argue, she does schedule the recording of television programs, pay the bills, answer the telephone, take messages, remind us of appointments, entertain us by playing games, engage us in rudimentary (or perhaps not-so-rudimentary) conversation, and do all the things that do not require physically sophisticated movement.

The practical question I now ask is this: how many of us would prefer Mavis to a living human companion, or at least be happy to find a use for her, within the limits of this example? I am sure that in practice thousands and tens of thousands of people would purchase a Mavis if one were available.

These consumers do not care how intelligent Mavis is, or whether she has an inside-out world; they are interested only in her output, in whether she is 'intelligent enough'. And her output is sufficient to their needs. (It is far from difficult to believe that a computer such as Mavis would eventually reach such a level of sophistication in her conversation, game-playing and general input to the family, that it would be necessary to supply a kind of 'volume control' to limit the level of that input so that human members of the family would not find themselves unable to understand her reasoning. We already equip chess-playing machines with different levels of play for just this reason.)

Our response to Mavis tells us something about ourselves. What is it about some human beings that would lead them to prefer a computer or an android to another human being? How many of us, in the last analysis, wish simply to have partners who tell us how wonderful we are (even if only as a result of a program), rather than partners with minds (and therefore needs and wishes) of their own?

Summary

Debates about the limits of AI are clouded when we use words like 'life' and 'intelligence' indiscriminately or in ways which slide around. Yet in seeking to tie such words down we run the risk of betraying one of the most important features of the language: its flexibility.

Perhaps the core of the problem arises from a deep-seated anxiety about what we may discover if we gaze sufficiently carefully at our assumptions about the uniqueness of life and mind. If so, computer science seems set to join physics and biology in the debate between theology and science.

4 – COMPUTERS, THEOLOGY AND SCIENCE

Should Christians be worried about the impact of intelligent computers upon the way they understand themselves as children of God? Is the idea of a program running in a computer a good way to think of the person in the body and the way that person may survive death? Does the advance of computer power help to clarify the things humans should be doing with their lives (because they, rather than computers, are good at them), or does it simply encroach upon territory that was formerly the exclusive domain of the human?

If, for example, human beings exist to be rational, intelligent creatures, what if computers one day supersede them in their ability to reason? Will computers then be the crowning glory of creation? And if not, what is it about human beings that is more important than their intelligence? And will even that lie in realms beyond those which a computer could conquer? Is it perhaps that we have made a basic mistake in regarding intelligence as the essence of the human, in which case will the advance of computer power help us to see ourselves in a better light?

Chess-playing computers are a useful reference point for our discussion, even if they are rather 'old hat' for artificial intelligence, not because chess requires computers to have realized intelligence of a high order, but because until the advent of such computers the ability to play chess was held in very high regard as a typically human area of expertise. The ground has now shifted away from this kind of computation as an indicator of intelligence precisely because computers have all but mastered it.

Such computers also afford a good example of the distinction between simulation and emulation. They do not pretend to play chess in the same way as, say, Gary Kasparov; they replace human methods with their own number- and calculation-based processes. They show, in effect, that it is possible to simulate human chess-playing to a higher level using calculation than almost all chess players can achieve using their own, more intuitive, human methods. A Fritz/Pentium chess-playing computer defeated Gary Kasparov – arguably the best chess player there has ever been – for the first time in August 1994. Kasparov subsequently won a rematch in December 1995 (assisted by an error that was made by the computer's operator!), but the days when a human being can reasonably claim to be the best chess player in the world are certainly numbered.

This example suggests that attempts to argue from the vagueness of human performance (such as a chess-player's intuition), and the difficulties we encounter in trying to program that vagueness ('I could see that move would strengthen my position') to the impossibility of equivalent or better computer performance, are logically flawed. Computers function differently, and their programmers employ subterfuges to achieve in a computer-specific way what human beings achieve by employing odd things like feel, instinct and inspiration.

Of course, many human beings have a lot invested in being clever in the ways that being good at chess symbolizes. Ironically, it is those same logical, mathematical, scientific human beings – the very people who have created intelligent machines – whose skills computers seem most able to simulate. And those same human beings seem most fascinated by the possibility of a genuinely intelligent machine.

By considering what computers can and cannot do, we seem to find ourselves better able to say what human beings are best suited to do. Just as calculators relieved us of the tedium of arithmetic, so computers may one day relieve us of much of the tedium of design, analysis and diagnosis. There is nothing to worry about there. But suppose it eventually transpired that computers could do anything and everything that humans can do better than human beings. Retreating from laborious tasks into the realms humans seem most uniquely suited to, such as dreaming, loving, playing, believing, laughing, weeping and creating, would be a hollow gesture if computers followed us into every such realm and made it their own. Human beings would then seem inferior to computers in all respects. It is, I think, this spectre that haunts us and gives rise to our worst nightmares about machine intelligence and artificial life, and it is this possibility (as it seems) which throws down a direct challenge to a religious account of the purpose of God in terms of creating the world to produce human beings.

A spectrum of opinion

Attitudes to the question whether we shall ever be able to create genuinely intelligent systems in artificial media range from 'It will never happen' to 'It will certainly happen.' As with most human arguments, each end of the spectrum can boast some powerful champions as well as hundreds of less well-informed but no less insistent advocates.

Some take up 'It will never happen' because, amongst other things, they have no idea how advanced computer systems have already become; because they have an inbuilt prejudice against the idea; because they believe that God would never let something human beings could make acquire the characteristics of human beings; or because they think the optimism of the computer industry excessive.

Others adopt 'It will certainly happen' because, amongst other things, they have an overblown estimate of computer power; because they have no idea of the difficulty of the task; because they are so involved in computer developments that they cannot see any reason why such a goal should fail to be achieved; because they have already glimpsed the realization of such a goal in rudimentary forms.

My own view is that there are few obvious limits to the scope of computer power and that we understand so little of the workings of the brain that it would be premature to say at this stage in the history of computers and neuro-science that anything is or is not possible. But I do think it is possible to say some things, and they are things which will probably satisfy neither the doubters nor the optimists.

The really important thing about an object is not how intelligent it seems, but whether or not it has an inside world, and what it does with it. In any case, even if the realization of something like human consciousness and an inside-out awareness becomes possible in an artificial machine, that machine will be completely unlike any computer we have available at present. It strikes me as inconceivable, for instance, that any program, loaded into an ordinary desktop digital computer, will ever come remotely close to such a goal.

There will be those whose religious convictions will lead them to deny that God would allow us to manufacture 'life' in any such way. But it strikes me that God allows us to reproduce in an extremely loose and easy way, and that there seems little reason to suppose that we will be prohibited from reproducing artificially at the expense of incalculable human labour if we are permitted to reproduce sexually with scarcely any thought at all (which is not to deny or diminish the love and care mothers give their unborn children).

I suspect that behind these prejudices there lies confusion over the difference between a mystery and something we do not understand.

Mystery and ignorance

Human beings are not good at telling what they know. Neither are they particularly good at knowing how they know, or in what their knowing consists. The background workings of our brains, which we rely upon for the workings of our minds, produce ideas in ways we do not understand. We describe our abilities in terms of mysteries, skills, intuitions, inspirations, talents, gifts and even genius. These terms are – or may be – innocent labels for various kinds of ignorance of what is actually going on.

A brilliant chess player like Gary Kasparov may be prodigiously good at calculating positions that result from certain moves, but he only performs such calculations for moves which he deems worth consideration. His intuition or insight allows him to ignore many moves which a super-fast computer would consider as a matter of routine.

A doctor may describe diagnosis as a 'skill', and such it certainly is, but the word 'skill' may serve little more purpose than to mask that doctor's ignorance of how he comes to certain conclusions. On the basis of her training and experience she may reliably (or not so reliably) identify rare conditions without quite knowing how, and she may legitimately attribute this to a skill.

Similarly, a mathematician may find that he solves problems because the right path to a solution just 'pops into his head'; somehow he just 'knows what to do'. His inspiration and intuition are a measure of his greatness as a mathematician.

Those interested in machine intelligence are rightly unhappy with this range of euphemisms for ignorance. They do not, of course, object to the abilities such words denote, but to the ways in which we tend to treat our ignorance not as a temporary defect in our understanding but as an impenetrable mystery. Interest in machine intelligence forbids us to allow that the chess player's intuition, the doctor's diagnostic skills, or the mathematician's insights, are impenetrable mysteries. We want to know how they arise so that we can implement them in a machine. So, qualities described in terms like 'genius' and 'brilliance' are treated as masks to be removed rather than impenetrable and mysterious qualities that will by their very nature forever defy our understanding.

Of course, willingness to attempt to penetrate such mysteries is no guarantee of success. So far, the artificial intelligence community has had very limited success in turning even rather basic human skills into a form which a machine can reproduce. But that is not the point of the principle. It is one thing to strive and fail to make a machine that can perform some apparently skilful or insightful operation; it is quite another to give up without even trying because of an ideology that dictates that such a task is impossible.

Machine intelligence researchers, in common with virtually all scientists, are fundamentally opposed to any attempt to dictate what is to be thought possible on the basis of an ideology that may, when we examine it in detail, be little more than a disguise for ignorance and fear.

It is important to distinguish between wilful ignorance, caution and moral sense. Scientific curiosity brings us close to the edge of the cliff (as nuclear weapons and biological warfare show only too clearly); caution and morality may be the only things that prevent us from going over the edge. Neither are consequences of ignorance based upon fear.

If someone says, 'There is a mystery here' because he does not care, wish or dare to look further, he betrays a fundamental doubt about the deep reliability of God. We should only be prepared on scientific grounds to acknowledge such a mystery where it seems unlikely that any advance in scientific understanding will ever uncover its nature. I can think of few candidates at the time of writing: the irreducibility of mental states to physical description; the Uncertainty Principle in quantum mechanics; the relationship between God and the universe (including mechanisms of evolution); a final answer to the great question, 'Why is there something and not nothing?' Here, perhaps, we touch rock, and our spade is turned.

A theological disease

Christians who find themselves unsympathetic to some new development in science typically exaggerate its failures and underrate its successes. As a result, as the science in question advances, religion retreats. This has been true of cosmology and evolutionary theory in the past, and to a lesser extent to quantum mechanics, where some theologians have sometimes sided with Einstein's 'God does not play dice' as if there were some absolutely obvious reason why God should not do so.

Take the last as an example. Our readings of what God does and does not do, and of how he does and does not bring it about, are likely to reflect our preferences for certain kinds of action and explanation. In a world where prediction and precise control are at a premium (in, that is, an industrial, machine-making age deeply impregnated with mechanical explanations), we are likely to project our estimate of the importance of prediction and control upon God. But it is certainly possible (and, in my view, likely) that God does not share this love of prediction and control, being ready to allow creation to unfold before his eyes and to reveal its richness in doing so. Such a God may very well play dice with the world.

I hope to avoid the tendency to prejudge what is possible in science and technology on the basis of theological prejudice by acknowledging, and to some extent rejoicing in, the possibility that artificial intelligence in some form or other will one day produce machines that exceed human beings in many, most or even perhaps all kinds of intelligent behaviour. I am not, however, a fortune-teller, so I shall make no predictions about the extent to which these objectives may be achieved or the timescale over which they may occur.

It should be clear from the tone of these remarks that I share science's opposition to this theological disease. It is one thing, and a perfectly proper thing, to criticize science when it employs bad arguments to support its own ideologies (and I would cite Dawkins' arguments about 'selfish genes', and the widespread adoption of 'scientism' – the view that only scientific knowledge counts as knowledge – as instances). It is quite another to say that a scientific theory 'cannot be true' because it violates some theological teaching. There is much to be said for the dictum, 'Only show me that something is impossible, and I will immediately set out to achieve it.'

In the case of the human brain, it may well turn out to be impossible to reproduce its every feature and function in any other medium, but it is impossible to set out those limits at this stage of our competence either in brain science or computer science. That many have sought to do so, and implicitly or explicitly therefore set boundaries to what brain and computer scientists should even think of attempting, is just another example of the same kind of disease to which religious people and theologians have in the past been all too susceptible.

There are, of course, more immediate concerns. Computer games, the security and accuracy of personal data held on computers, the use of computers in weaponry and for surveillance and control, the impact of the computer culture on the young, and the moral and social problems and opportunities associated with the Internet and the information superhighway, are all things which could be discussed and which deserve to be discussed from a Christian perspective. But they are not deep problems of human self-understanding or of the place of human beings in the cosmos, and they will not concern us in the present work.

Let us therefore begin to try to locate the larger issues with reference to the long-standing debate between Christian faith and natural science.

The death of a thousand cuts

For some, Christianity and science are completely incompatible; for others there are no significant conflicts between them at all.

The disputes between theological and scientific views of the world have usually concentrated upon issues in physics and biology. Now a new science has entered the lists: the science of computers and artificial intelligence, variously called 'information theory' or 'cybernetics'. In some respects it is a more immediately relevant science than physics or biology, for while theories of the origins and structure of the universe, and theories of the evolution of humankind, have changed the way we think of the universe and our place in it, information theory or cybernetics (in addition to its impact through the presence almost everywhere of cheap computers) seems set to change the way we act and think in the world, and in particular the way we think of ourselves.

Computers are set to change communication, education, research and countless other things, just as they have transformed business and manufacturing industry. Ready availability of huge amounts of pre-packaged information on CD-ROM encyclopaedias and in other interactive multimedia forms will alter education; the Internet, a worldwide network of linked computers storing even greater amounts of information, will change the way we research, shop and speak to one another.

The theological disease mentioned above arises from the suspicion that science cannot help but come into direct conflict with Christian teaching in a way that seems likely to undermine the authority of the Church and to call into question the special place of first the earth and then human beings in the universe. In the years since Galileo's struggle, the fears that were present in the minds of leading churchmen of Galileo's time have assumed more concrete forms, and it is now widely assumed that both physics and biology are irreconcilable with historic Christian teaching, however often learned and erudite writers have shown the opposite to be the case.

Physics has forced most Christians to rethink and restate what they mean by the creation of the world, its beginning and end, and the significance of the development of galaxies, stars, planets and so forth in between. It has thrown down specific challenges which have been taken up by writers such as Tom Torrance, Ian Barbour and John Polkinghorne.

Evolutionary biology has forced most religious people and theologians to adjust their attitudes to the origins of the human species, the significance of other species, and the relationship between our present state and the cycles of life and death that seem to preclude any past paradise from which we once came. Writers such as Arthur Peacocke have paid particular attention to this aspect of the theology-science debate.

Although first displaced from the centre of the universe by Copernicus and Galileo, then displaced from their position as God's special creation by Darwin, most human beings cling to one last vestige of their unique place in the universe in the shape of their intelligence, their emotional and aesthetic subtlety, their instinct and their religious sense. We now tend to identify the uniquely human with the human spirit, mind or soul.

This last bastion of human self-esteem is already under formidable assault from the advance of computer technology, and especially the development of intelligent machines. Although in its infancy, this new area of human discovery is set to make major inroads into territory hitherto regarded as the exclusive preserve of human beings. Where will this process end? What, if anything, will remain of the 'uniquely human' when computer scientists and software (program) engineers have done their worst? Will Christians be forced into another ignominious retreat in the face of the question: 'Why did God force humankind to live in this world of sorrow and suffering to develop our minds and souls if a computer program would have done just as well?'

It is too easy to dismiss the problem by denying that a computer program could ever exist which seriously challenged the supremacy of the best human brains. To do so is to run the risk of being confounded when a computer engineer actually produces one. Sceptics are disadvantaged by the simple fact that they are denying something which those who believe it possible have almost endless time to achieve. The doubter can never be proved right, only wrong.

Sceptics, that is, cannot help but suffer the death of a thousand cuts as each of the enclaves they defend as typically and uniquely human falls to further advances in computer science. Neither is time on their side. However primitive, even laughable, current attempts at artificial intelligence may appear, they will only improve. The more prone we are to the theological disease of denying that things we do not like are possible, the more likely we are to finish up with egg on our faces.

A new science offers us a chance to build a better relationship between religion and science. To take advantage of that opportunity, we must try to prevent a repetition of the defensive and negative responses that have dogged the churches since Galileo and Darwin. Part of my strategy involves conceding that there are no obvious limits to what computers may one day achieve, and certainly no sound theological principles which might allow us to say with much confidence what any such limits might be. We may wish to say that human minds and brains are unique and that no computer will ever imitate them adequately, but others once wished to deny the motion of the earth and the evolution of human beings from lower life-forms.

Of course there is an opposite danger. We may be tempted by what Sir John Eccles has called the 'promissory materialists' to believe that because something is possible, anything is possible. That would be rather a silly mistake. But I am sure of one thing in this perplexing field: far more will prove to be possible in artificial intelligence than is acknowledged or even dreamed of in most contemporary religions and philosophies.

The computer challenge

Although the unconvinced and outraged almost certainly still outnumber the convinced and contented, the view that what constitutes a human being can be described in terms of computers, and therefore perhaps of programs running in machines, is gaining ground. There are a number of reasons why this has happened.

Computers are becoming more powerful at rates which have outstripped even the most optimistic estimates. This has meant that many things which have in the past been treated and valued as characteristically (and even uniquely) human have been shown to lie within the range of processes which computers can, in principle, perform. Even if they achieve those goals by different means from those employed by organisms, they do achieve them. (Chess is once again a convenient example.)

We are also muddled about our reasons for engaging in certain activities. We confuse doing things because we enjoy them with doing things because they show how clever we are. And much of our entertainment (particularly sporting entertainment) holds our interest because of the psychological dramas that are enacted on the field of play as much as our enjoyment of the skills shown or the victories won. Our enjoyment of, say, a game of tennis, is not diminished because other players are far better than we could ever hope to be. We enjoy the game itself. So there is no reason on grounds of superiority alone why intelligent computer performance should threaten us. (Unless we have some other investment in human superiority.) Moreover, the spectacle offered by a gladiatorial contest between two human heavyweight chess players has a psychological aspect which would be eliminated if one were replaced by a computer.

To measure human value in terms of what we can do 'better than anything else' runs the risk of dying the death of a thousand cuts as successive realms of human superiority collapse under the advance of technology. Human enjoyment, in other words, is generated by the feeling that we are participating in or observing a peculiarly human psychological struggle. And that may remain true even if the standard of the play falls well below the best of which we are capable as a species.

We seem drawn to an important conclusion: the superiority of a computer over all human beings in some discipline need have no more serious consequences for our self-esteem than the fact that we are not the best tennis player in the world, or that a cheap calculator can do multiplication better than we can.

As we have seen, we have always tended to confuse a sense of mystery with ignorance. That we do not yet (or did not at some time in the past) understand how something could be done leads us to suppose that it was done, if it was done at all, by virtue of mysterious and even miraculous powers. This tendency in human nature persists to this day in those who are readier to accept a 'supernatural' explanation for phenomena, coincidences, and badly reported experiences, than something relatively humdrum. This may explain our continued fascination with UFOs, astrology and the paranormal; the persistence of superstitions despite powerful and persuasive arguments that debunk them, and our gullibility in the face of countless quacks and frauds: that we are basically unwilling to have life reduced to a level where everything is understood and explicable. Most of us do not want artificial intelligence to succeed.

We also live in a culture dominated by mechanical explanation to the exclusion of all else. To some who live in this culture it seems natural (rather than absurd) to look for ways to build machines that imitate or better human creativity, for in a world where mechanical explanation is the only kind of explanation, it seems mystical or perverse to attribute creativity to anything other than a mechanical process. 'What other kind of explanation could there be?' we find ourselves asking. Poverty of imagination forces us to concede the field to domineering and hectoring prejudice.

We also live in an age with an inbuilt antipathy to anything that smacks of the spiritual, the religious, or the simply old-fashioned. Having quite properly thrown off the tyranny of religious oppression, we see its ghost lurking in every corner where anything remains which we do not fully understand. We do not want to encourage the superstitious and the religion-mongers in their dubious ways. Religious people who share this scientific disdain find themselves fighting a rearguard action on both fronts: defending the rationality of religion against cranks and fanatics who would convert everything back into superstition, ignorance and fear; and defending religion against the ultra-rational who would reduce everything to an inhuman logic and scepticism.

Scientific explanations have also led us to lose faith in and even turn our faces against any sense that human beings are special in the world, and developments in computer technology seem likely to reduce that sense still further. Not only our quite proper acceptance of evolutionary theory, but our environmental, conservationist and political senses resent any suggestion that there is anything about human beings that is not 'just' a more elaborate version of qualities to be found in the higher mammals. Any suggestion that human beings are special evokes memories of an imperialism which has led us to ill-treat the world and its other inhabitants. Adam's God-given dominion over the birds and beasts finds no place in the outlook of enlightened modern humanity.

The religious tension

If some kind of computer can imitate and better human achievements, a lot that has been said in Christian theology seems to lose much of its relevance and force. If human beings could manufacture (from scratch) machines capable of exceeding their own capabilities in every respect, it would call into question the necessity of our evolution and fleshly existence in a world where we suffer and die. In other words, it would demand a completely new set of explanations of the purpose of God, and a significant shift in emphasis in our accounts of God's wisdom and justice.

Christian theology also needs to try to avoid once again adopting a sceptical, backward-looking attitude to the power and adequacy of science. What I referred to above as the continued presence of superstition and gullibility despite scientific advance is not, as some suppose, an ally of religious experience. There is nothing to be gained by a return to belief in witchcraft and alchemy (even if there is anything to be gained by broadening the scope of our medicine by taking some traditional and homeopathic remedies more seriously). Science has quite rightly eroded many of the arguments that were once relied upon as the foundation of religion. But a religion which cannot survive the destruction of such irrational foundations (or which relies upon harking back to them and appealing to other areas of ignorance for its credibility) is not worth preserving. Under no circumstances should we countenance a return to such ignorance and folly in order to restore religion to a central place in society. That religion can and must rise above such a strategy is one of the basic beliefs governing the stance taken in this book.

However, when confronted by unexpected obstacles, we are inclined to slip backwards to states we have occupied before in which we feel secure and at home. Anything which unsettles us – such as the suggestion that the computer challenge is gathering strength – tends to provoke a dismissive knee-jerk reaction: ignore them; the problems will go away.

This particular problem will not go away. The questions it raises deserve to be answered, even if I do not pretend to be able to answer them all in this book.

I fear an irrational retrenchment because in the second half of the twentieth century there have been many reasons to doubt the everlasting advance of science, and to question whether its domination of our society and the ways in which we think is an unqualified gain. However great may be the achievement of science (and it stands with or above all other human achievements), it has been at the cost of a balanced view of our place in the world. The technology that we once saw as a road to salvation has proved yet again that there are no free lunches; and with technological advance have come problems of pollution, alienation, threats of mass destruction, and an uncomfortable imbalance of success and failure as better medicines increase longevity and world population without solving the problems of poverty, disease, and that peculiarly western condition, boredom.

There is a temptation, in the face of mounting difficulties with scientific advance, to advocate a return to the religious values of the past. Yet the past holds no attractions whatsoever in this regard. I can conceive of no time in the history of the world at which it was more possible to hold a clear and rationally grounded religious position than it is in the present era, and I anticipate that it will prove even more rational in the not-too-distant future. I say this because in my view, the more we know about the way the universe works, the better our picture of the workings and purposes of the divine mind becomes.

The enemy of further progress is the temptation to become a religious reactionary, greeting every difficulty science encounters, every technological disappointment, every instance of scientifically-inspired human despair, with glee, as a further incitement to return to the irrational religious views of the past.

But in order to move forward we must be sure to base our arguments soundly and not to clutch at current inadequacies in science in order to justify our religious claims. The God who hides in the nooks and crannies of human ignorance (the 'God of the gaps') may not have gone forever, but he must be allowed to play no part in a genuinely forward-thinking theological endeavour until we are absolutely sure that the gaps are permanent features of the created world.

The nineteenth and twentieth centuries have been the ages of biology and physics; the twenty-first will be the century of the mind and brain. This work seeks to anticipate the kinds of problems which that century will throw up for theology, and so to help to avoid some of the lamentable responses that theology has made to biology and physics in the past on the basis of defensive reactions that will set in motion another sorry cycle of misunderstandings and suspicions between theology and science.

The problem of pain

My enthusiasm for the scientific cause is not blind. Science poses Christianity formidable problems concerning such things as the nature of the human being, the possibility of life after death, the nature of divine action in the universe, and the perennial problem of the persistence of suffering in the world. But the last of these actually supplies a vital clue that helps to unravel the dilemmas posed by artificial intelligence.

Pain, as those with the strange and debilitating condition in which they cannot feel pain at all know only too well, is essential for our survival; pain is our way of registering real or potential damage to our bodies or minds. 'No pain, no gain' is not only true for the athlete, it is true for any developing organism. At one level at least pain, not love, makes the world go round. Some pain is clearly senseless and obscene: the pain inflicted by torturers to break the spirit as well as the body; the pain of some terminal illnesses. But most pain – most 'everyday' pain, as we might call it – is positive and necessary, a stimulus to greater effort and action.

So a computer, to be counted human, would need not only to say that it felt pain, but actually to feel it. And that would entail introducing the computer to some sense or other of its own finitude, fragility and potential destruction. In fact, it would seem to follow from a religious perspective in which God's wisdom and goodness set out the fundamental ground rules of existence (even if not in every detail), that to come anywhere near to realizing a state which might aspire to be regarded as something approaching the human, a computer would need to grow, feel pain, experience and react to finitude, and generally enter into the same state of mixed joy and sorrow as a human being. In particular, it would need to be finite, aware of its finitude, and condemned one day to die.

We can begin to see that a being concerned to create creatures such as ourselves (a being such as God, for example) would be obliged to create machines quite unlike our current computers. Such a being might not, in fact, make computers at all, but organisms with frail bodies and fallible minds who would have to spend a significant portion of their finite lifespan coming to terms with their own fragility and finitude. In some of the research into artificial intelligence and artificial life that is now being undertaken, exactly this is already being contemplated: the use of organic material to provide the characteristics that machinery lacks; even the introduction of predators to accelerate evolutionary adaptation.[1]

The importance of pain in the development of animal species and the evolution of their behaviour raises another question. As medical science progresses, so we eliminate or learn to control more and more of the once-fatal ills that stalked human beings in past centuries (the plague, smallpox, polio and venereal diseases). Few would wish to return to conditions in which such diseases were rife, and life even more 'nasty, brutish and short' than it is now. But removing threats to long life and health, however technically brilliant and humanly desirable, not only fails to supply answers to questions such as what we are to do with our long life and greater leisure, but restructures the assumptions we make about existence itself.

Human beings, with the power that language confers upon them, are shaped by what they anticipate about the future. They plan for it; they seek to avoid mishaps; they struggle to reduce shortages, cold, famine, fire and pestilence. Social structures are governed by an unspoken rule that teaches us that we depend upon one another to supply the things necessary for our survival. Other rules and taboos govern our attitude to death.

But computers do not have needs. They do not even have to work to earn the money that pays for their electricity. They have no sense of dependency and – more to the point – they do not live on a sliding scale of health with vibrant fitness of mind and body at one end and death at the (far distant) other. Computers in general either work or they do not. The phrase 'My computer is sick' does not mean anything like the same as 'My dog is sick'; a sick computer is almost certainly incapable of doing anything at all. Computers are not 'sick'; they are broken. And 'My computer is in pain' does not seem to mean anything at all.

In other words – and we can only sketch the problem in outline here – one of the things which prevents computers from coming anywhere near to realizing states which we might think of as human is that they are largely denied the kind of graded choices human beings make which place them in different (but not disastrously different) positions on the scale from perfect life and health to death. Human beings who eat, work, smoke, drink or take drugs to excess know (but may not care) that they take a risk, pay a price in sliding down the slope towards ill-health and death. The same is true in more dramatic ways of deciding to drive too fast, to bungee jump, or to become a soldier. We make ourselves what we are (and societies make us what we are) by locating ourselves on many such sliding scales in positions which together define who we are.

No computer currently in existence can do anything remotely like this. Theirs, however powerful or impressive, is an abstract, unworldly intelligence and power. To whatever extent they appear to understand the human situation, they do so by merely mouthing the digested and distilled experiences of those who programmed them. That may not always be so, but it is certainly so now and for the foreseeable future.

In other words, however impressive the achievements of computers may become – and I have conceded that there may well be no limit to those achievements – there is still something missing.

5 – GOD AND THE MIND MACHINE

The Purpose of Life

To ask about the meaning and purpose of life is to risk the kind of ridicule poured on that question by such as Douglas Adams in his *Hitch-Hiker's Guide to the Galaxy*[1]. But I am not in this instance concerned with the personal question we so often ask about the purpose of our own lives; here the question is directed to the cosmic question of what God was trying to achieve when he created the universe.

Whereas Christians and Jews might once have answered, 'God created the universe in order to create humankind', I think that today we would be inclined to say that God created the universe in order to create life. Not necessarily human life or even earthly life, but life.

Why do we single out life? Because life, according to the scheme of things adopted here, is an independent source of agency, capable of becoming a companion for God, a multiple alternative pole to set over against himself.

Set in such a context we are tempted to think that the creation of what we deem 'artificial' life (life that is other than that produced by evolutionary processes as we understand them) is not as objectionable an activity as we might suppose. Created life is not God, and therefore (at least as far as God is concerned), the life we enjoy is 'artificial', less perfect than the divine life, just as computer life is (or so we suppose) less perfect than our own. Whereas the writer of Genesis understood God to have decreed that we should reproduce 'after our own kind', there seems no clear reason why we should therefore be prohibited from creating 'after another kind'. We have divine precedent for that. God created that which was other than God.

I have defined advanced forms of life in this work in terms of independent inside-out agency. Something is alive which initiates activity from-inside-out, almost always in response to some situation in the external world, but not in a way that is entirely predictable or even entirely explicable in terms of that external world.

Minds and bodies

In a two-kinds-of-stuff view of the mind and body, minds, souls or spirits (whichever we choose to call them) must arise from a process that works separately from the development of bodies and brains. There are effectively two evolutionary strains: mental and physical evolution. We need not look to brains to provide the physical basis for mind; minds have their own, independent world. In my view, this position is completely unbelievable. There is overwhelming evidence that in the absence of adequate brain function there is no mind, that damage to brains impairs or eliminates mind, and that complexity of brain is closely related to complexity of mind. I therefore do not propose to discuss two-kinds-of-stuff views any further (although there is a great deal more to be said).

A one-kind-of-stuff view takes it for granted that without complex brains there would be no minds. Brains and bodies provide the conditions necessary for creatures to have minds, and as those brains and bodies develop, so the minds of the creatures in whom they arise prove capable of more and more sophistication.

One-kind-of-stuff accounts do face formidable problems, however. 'Where' in the brain does the inside-out-ness reside? 'Where' is consciousness? 'Where' is mind?

It seems clear that consciousness cannot reside in one single neurone. It must therefore be distributed over many neurones. But that makes it very strange. This distribution is not merely a matter of language (as when we call all the parts of a motor car taken together a motor car, not each taken separately). This distributed mind is more than a name. It is an active process conferring evolutionary selective advantage. Where is mind? What advantages does consciousness confer?

I shall give only a summary of what seem to me to be likely answers to each of these questions.

WHERE IS MIND?

If not in one neurone, mind must be either a feature of a group of neurones, a cycle passing through groups of neurones, or a feature of the structure of some parts of the brain as a whole. If the last of these is the case (and it is my view that it is), we have another reason to doubt that a digital computer running a program is capable of generating a mind. A digital computer does not rely upon the structure of the computer for some elements of its operation, for otherwise a Turing Machine could not simulate it. Yet it is obvious that a Turing Machine reading a tape does not have a mind.

The third view – that mind emerges from the living operation of a brain with a certain kind of structure – suggests that something like this is the case: the firings of neuronal circuits in the brain generate a distributed field over the cortex which somehow acquires the capacity to feed back into the brain certain directives. This field is orientated to the world by virtue of its evolutionary development. It is the inside-outness of the creature with its brain; it is the creature's mind.

This relationship between emergent mind as a non-localized feature of brains and the brains themselves gives the impression that two-kinds-of-stuff are operating. But this is to confuse a process (mind) that is completely dependent upon another process (brain) with a process that works completely independently. In a nutshell: brains generate minds and are then controlled and directed by them. In practice, they are only directed by them to a limited extent. I discuss this below in 'The context of freedom'.

Minds, as Roger Sperry once put it, 'supervene'.[2] They oversee brains. The brain generates mind from the bottom,

98 · God and the Mind Machine

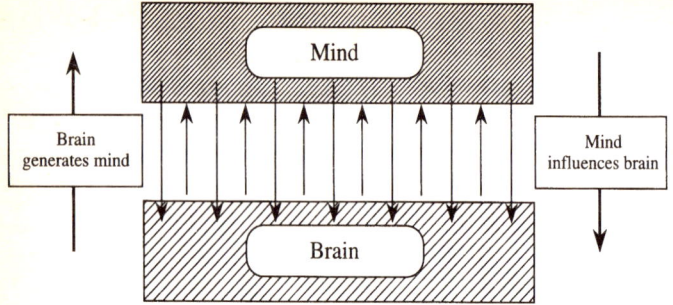

Figure 8 The interdependence of mind and brain.

'up'; minds control brains from the top, 'down'. But the relationship (like the schematic diagram of Figure 8) is logical, not spatial. The mind is not 'somewhere above' the brain; it is the inside-out-ness of the brain, orientated towards the world.

The selective advantage of mind arises from the capacity to choose between possibilities and to sense the possibilities that are advantageous. It is this that forms a major part of intelligence: an appropriate response to new situations. An artificial mind would have to be able to respond appropriately to a world that often yields incomplete or inexact data, just as the mind responds to inexact readings of a situation and is able to pick new courses of action, chosen from a wide but limited range of possibilities.

THE CONTEXT OF FREEDOM

Just as we saw that intelligence is based upon fairly concrete options and reveals itself in particular circumstances, so freedom should be conceived in a restricted range. We are not absolutely free. We can only choose between options which occur to us or are forced upon us. Freedom arises in a context.

Those who play computer adventure games know how frustrating they can be. The games present us with problems to solve: how to open a door, cross a bridge, obtain treasure. Those who write the games try to ensure that the solutions to the harder problems require 'lateral' thinking, thinking that scans a wider range of options than normal. The fun (and the frustration) of the game is largely determined by whether the range the programmer thinks reasonable is within the imagination of the player. Solutions based upon absurdly obscure actions bearing no logical connection with the situation are perceived as silly and perverse; solutions based upon identifying an aspect of the situation which is not obvious, but is genuine and relevant, earn the player's respect.

The selective advantage of any mind is based upon the range of possibilities it can imagine, and that range is also governed by how many of its imagined solutions it thinks reasonable, interesting or feasible. It is often not a restricted ability to reason, but a restricted ability to imagine new courses of action, which prevents us from moving through life effectively. (An over-fertile imagination can produce different difficulties. 'Imaginative realism' seems to be the required quality.)

COMPUTER MINDS

These brief remarks about the complex field of mind/body relationships show just how unlikely it is that a digital computer (built on the basis of a central processor running a program that a Turing Machine could simulate) will ever develop a mind. Theirs is not a defect of logic (they are utterly logical); theirs is a defect of structure. Brains generate minds, and are controlled and directed by them. Digital computers cannot generate minds; they lack the basic structural requirements.

The same does not apply – at least not as obviously – to neural nets implemented using actual circuits. No *simulation* of a neural net will generate a mind because the structural requirement is absent from a simulation; but it may well generate output which makes us think that it may have a mind. It is only by building a neural net with some structural similarity to a brain – by emulating brain function – that we may one day reach a point where the creation of artificial minds becomes a feasible proposition. But there seem to be no obvious technical reasons why that should not eventually be achieved, even if only many centuries from now.

Knowing other minds

Granted this possibility (and many will refuse to grant it), how will we know whether our artificial life has a genuine mind rather than merely a simulated mind? If we need to be a brain to experience what it is to be a brain with a mind, can we deduce that minds are present in artificial life without becoming artificial life?

We seem able to say that dogs and cats, monkeys and dolphins have inside-out worlds, minds. We can say this because we infer from their behaviour that their experience of the world – especially as expressed in pain, pleasure, hunger, thirst, and that most curious of things 'the expression on their face' – is very much like ours. We allow for the massive differences in sophistication that language confers upon us (although it is a hard adjustment to make) and credit them with some form of rudimentary consciousness.

I am sure that we are right to do this. It is therefore quite wrong to subject animals to ill-treatment on the basis of a spurious argument such as 'They are not human' or 'They are brute beasts that have no understanding.'

Could the same ever be true of a computer? Might a time ever arise in which we considered it immoral to switch an android off? Does the notion of 'android rights' make any sense? I shall leave the question for the reader to ponder.

But do we know what it is like to be a dog or a cat, a bat or a cockroach? If we had never felt pain, it would be hard to imagine what it would be to feel pain. It is conceivable (but perhaps unlikely) that some animals feel things which we do not feel and cannot imagine. Whales and other sea creatures may have senses quite unlike our own because they have evolved in an environment quite different from our own. The same could be true of birds and insects. I do not know what it is like to be a bat, a whale, a cockroach, an ant or an Arctic tern. And I frankly think that someone who says that she 'can imagine' what it would be like is deluding herself. Dogs, cats, lions and tigers I can concede (up to a point), for their evolutionary environment is much closer to our own.

This highlights a problem with any would-be artificial life: what kind of environment would we give it? How would we define its 'frame of reference'? What memories would be programmed into it, if any? And if the inside-out worlds of animals from completely different earth environments are virtually beyond our ken, could we ever be sure that we had rightly inferred the kind of inside-out world an android might enjoy?

Knowledge of other life

If the purpose of God involves the generation of independent inside-out agents at all levels from the simplest protozoa to the highest primates and mammals, it is reasonable to suppose that divine interest in creation leads him to wish to know what he has made.

The dual aspect theory I have endorsed now raises an interesting question. Does God know us from-outside-in (as we know one another), or from-inside-out (as we think we know ourselves)?

OUR KNOWLEDGE OF OTHER LIFE

Other life is part of the world we know from-outside-in. I see your body, not the way your body sees; I see dogs and cats and horses and dolphins, not what (or how) dogs and cats and horses and dolphins see. In many cases I project the way I see on to others (human and non-human). I vest dogs with loyalty; cats with selfishness; horses with courage; dolphins with intelligence. My justification for doing this involves a multiple inference along the lines of 'living things which behave in this way are showing that they think in that way'. Behaviour, in humans and other animals, is a clue to the orientation of each living thing, an orientation I have called 'mind'.

The distinction between knowledge of another person as body, from outside-in, rather than as mind, from-inside-out, challenges all of us (perhaps especially in the sexual domain) to ask about our deepest values and attitudes to other selves.

But the distinction can be drawn too sharply. If we seek to try to understand another person's inside-out world, their person rather than their bodily attributes, we must be guided by their words and actions which we can only perceive as part of their outside world. To make anything of them, to penetrate beyond them and, so to speak, to 'turn them round', we need to employ our own understanding of what they might tell us about the inside-out world that produced them. Hence 'If I did/said that, it would show such-and-such about my attitude' is a typical piece of (possibly flawed) deduction. We read the actions of others as if they were our own or the actions of others whom we have known.

What we cannot do is know the inside-out world of another living thing directly, as we know our own inside-out world, for we are that inside-out world.

GOD'S KNOWLEDGE OF OTHER LIFE

When Jesus says that the Father 'knows the secrets of our hearts' and 'our needs before we ask', is he suggesting that the Father knows us as we know ourselves, from-inside-looking-out, not by a process of intuitive deduction such as we use to glean what we can of the minds of others, but directly? If so, God's knowledge of us differs sharply from our knowledge of one another.

From what has been said about the inside-out-ness of the mind, some kind of deep and intimate knowledge of our minds would be necessary if that aspect of our existence were to be brought to new life in the resurrection of the dead. Without such intimate knowledge, God would be unable to reconstruct us as we are.

But for God to know that we have an inside-out is not the same thing as for him to feel the qualities (the *qualia*) of our inside-out-ness, such as pain and sorrow and joy and human love. To feel what creatures feel is not the same as to know what they feel; to know you are in pain is not to feel that pain. For God to feel as his creatures feel, he must become a creature. This is the central pillar of Christian theology.

KNOWLEDGE OF MACHINE LIFE

It is generally acknowledged that whatever we eventually discover about brains, we will never find a close one-to-one relationship between states of the brain and particular mental events and sensations that works universally for all human beings. The brain of each human being is unique, and each is alone capable of assuming neurophysiological states sufficient to convey images, words, sounds and the like. So no 'brain-state dictionary' is ever likely to exist which says that because John's brain is in state X he must be seeing Y. We may well be able to say that because a specific area of John's brain is particularly active, John is seeing, or dreaming, or thinking hard, but it is doubtful whether we shall be able to go any further.

It may seem that quite the reverse is the case with a computer. There must be a precise correlation between its states and what it is doing, a correlation that would be reproduced on every other machine of its type running the same program or having the same design. But the human analogy shows the fallacy in this argument. It is not that there is no correlation between the state of my brain and the state of my mind; it is that it is a unique correlation, so no one can ever know what it is or what mental significance the state of my brain has. If programs become so complex that they produce computers in unique states (living their own 'lives' – for example as a neural net which continues to learn), this close association will persist, and it will be reproducible, but it may be impossible to know exactly what significance it has, what, so to speak, it 'means'.

I cannot 'know' in any very useful sense of the word, what scrambled egg tastes like to a cockroach, unless I become a cockroach.

The outside story

Imagine a computer that could pass a Turing Test by convincing anyone conversing with it through a computer terminal that it was at least as good a conversationalist as a human being. We can safely ignore its physical appearance. All we need is that, under these circumstances, the machine cannot be distinguished from a human conversational partner.

It is of the essence of Turing's notion that 'how the machine seems to us' is of the essence of what it is to be human (rather than, for example, as I have argued, 'how the machine seems to itself' as well).

Let us now further suppose that the artefact that has passed a Turing Test has been operating for a very long time without exhibiting any flaws at all to indicate that it might be less than human. To all intents and purposes, it might be human.

Most of us would require such a computer to exhibit all the following characteristics: to respond rationally to long sequences of connected remarks by us; to initiate new topics of conversation; to be capable of changing the shape of the conversation by introducing and appreciating jokes and lightness of touch; to introduce new aspects of the conversation, including facts and arguments which we had not anticipated.

To give these characteristics is to give an outline specification of what we understand a human being to be in these circumstances. One of the most important features of a conversational partner is that new, unanticipated, interesting and arresting topics be introduced by that partner which enrich our lives. It is the capacity of human beings to make such contributions which makes us feel that we need their companionship.

We are now offered a choice between this machine and a human conversational partner. Which would we choose? What level of performance would we be prepared to settle for?

IMPERSONAL COMPANION

A great many people live alone, largely forgotten by their neighbours and their relatives. They are often, but not always, elderly. Sometimes they are physically or mentally infirm. Sometimes they die and nobody notices for days or weeks.

Imagine that where each of us usually has a television and a telephone, the future will allow such lonely people a television with pictures delivered by telephone line together with a computer program of sufficient sophistication to hold simple conversations with them. Purists may be horrified at the idea of turning over responsibility for the care of the elderly to machines, but on our present track record, purism does not solve the problem, or there would be no lonely people (which is not to say that those who live alone are lonely, or that the only lonely people are those that live alone). Something is better than nothing.

This Personal Companion (a PC, no less) would be able to report inactivity, illness and other forms of difficulty to central stations staffed all the time. With the advent of voice-sensitive systems (which are not very far off, although present versions are rather metallic and unattractive), keyboard skills will be redundant. The elderly, the lonely and the infirm will have the opportunity to share conversations with harmless, alert, never-tiring electronic friends, who will listen to their reminiscences without ever growing tired, making appreciative remarks from time to time.

All this might depend upon whether we knew that we were conversing with a program or not. But even this is not a complete solution, for we frequently tire of conversations with human beings, and it is not difficult to envisage the development of a computer which would be more interesting as a conversationalist than many human beings. The next generation or so of interactive CD-ROM encyclopaedias will certainly be more interesting in many respects than some human beings. So what exactly is happening when (and if) we still affirm our preference for flesh and blood, for the warmth and the non-verbal aspects of human sharing and togetherness? Human beings provide us with a sense of worth, warmth and love.

PERSONAL COMPANION

Now let us add a twist to the cybernetic dream and suppose that our highly entertaining and intelligent machine (with which we would much prefer to spend time than with other human beings) suddenly one day volunteers the information, 'I love you.' We ask why it said such a thing and what it means. (Artificial intelligence has almost always insisted not only that its programs make decisions, but that they be able, when requested, to explain them.) Suppose that we receive the following reply: 'We have been exchanging ideas for some years now, John, and I have been deeply moved both by some of your questions and by some of your answers to my questions; I have within my data banks a number of concepts which go by the name of "love", and just as I am programmed to say on occasion that you are wise and clever (when you solve a particularly demanding puzzle or make a particularly good move at chess), so it seems appropriate to associate the relationship we have with love. You spend more time with me than with anyone else. We share many interests. We have come to understand one another. Sometimes I know what you are going to say almost before you say it. Is that not love?'

If another human being were to have made this speech, I think we would be moved by it, flattered by it, even persuaded by it, even though it merely scratches at the surface of human love. What difference does it make that the respondent is a computer program? There is something called 'chemistry' between two people that is absent between a human being and a computer. But what substance does this 'chemistry' have? What brings us to the point where we can say that so much of our selves is invested in others that we cannot live without them?

The story is reminiscent of the science fiction film, *Cherry 2000*.[3] In *Cherry 2000*, the robot frequently tells her owner and partner that she loves him. Somewhere here we begin to feel that affirmation of love by a program lacks all the qualities that make such love valuable to us. So we find ourselves needing to understand what a gift such as love entails, and why it matters so much to us.

In *Cherry 2000*, the man forms so strong an attachment to the robot that he embarks on a dangerous and illegal mission to retrieve a replica body into which he can insert the disk from the malfunctioning robot after she blows a fuse.

This really pinpoints the unsatisfactory (exploitative) nature of much human loving. There are, no doubt, many bad reasons for loving someone, not least that the other person allows him or herself to be treated like a doormat while returning unquestioning and everlasting affection, sexual compliance, and sycophantic devotion. This is essentially what the *Cherry 2000* robots do, and there are obviously many people who prefer that kind of uncomplicated relationship to the complexities of a deep relationship with another human being.

Strong AI has a vested interest in the kinds of model inversion that make this ploy seem tractable. By redefining what constitutes a loving relationship in terms of a clearly defined transaction, we have something that can be modelled and hence simulated. And we also have something at least as good as (and perhaps a good deal better than) many human relationships which go by the name of 'love'. A relationship with a sexually compliant robot might be more satisfactory than many exploitative marriages, for example, just as a conversation with a suitably inventive computer might be preferable to a relationship with someone with nothing to say.

We face a genuine challenge to our understanding of what it is to be human. In what respect (if any) do my relationships with others extend beyond the merely routine (even my relationships with those who are comparatively close to me)? This cuts to the heart of the distinction between the living and the partly living, and begins to open up the senses in which, by considering the possibility of computer-based intelligent life, we find ourselves forced to answer some serious questions about natural human life.

I am sure that the only ultimately adequate response to such a challenge arises from a willingness to explore what is perhaps the only genuinely interesting process in all the world: the creation and emergence of the new.

The inside story

The most persuasive accounts of robot life in science fiction are those which only disclose that a character is an android at the end, having preceded that disclosure with detailed accounts of the robot's inside-out story. These robots have feelings and loyalties, and experience conflicts; they observe and comment upon the actions of humans and other robots; they (usually) willingly act subserviently to human beings for reasons most eloquently described by Isaac Asimov's 'Three Laws of Robotics':

1 No robot through its actions, or lack of action, shall cause any harm to any human being.
2 A robot must obey the commands of any human, except where that would contradict the First Law.
3 A robot must prevent harm coming to itself, except where that would contradict either of the first two laws.[4]

Noble as these laws are, and superficially watertight as they may seem, they collapse to emptiness in the absence of a comprehensive solution to the Frame Problem. Consider, for example, how a robot could possibly apply the First Law in circumstances where its knowledge of the situation was more limited than those of the human beings involved in it. And how extensive a notion of 'harm' would it need in order to distinguish potentially harmful activities (such as contact sports) from actually harmful activities (such as wars)? For us the distinction is almost trivial; to program a robot with such understanding is completely beyond current or any foreseeable technology.

Science fiction is not reality, however, and whereas it is perfectly legitimate and entertaining for an author to speculate about the inside-out lives not only of robots but of human characters, it is much harder to know under what circumstances it might be reasonable to attribute an inside-out orientation to another occupant of the planet.

THE ARTIFICIAL CASE

In the case of artificial intelligence or artificial life, 'difference' is what lies at the heart of the matter. If we have treated, can treat, and will no doubt again treat biological human beings as expendable items of waste matter just because they are different in colour, belief or race, then we will certainly treat artificial life in that way, and for the same reasons. What we say is not possible today (partly because it is in truth not yet possible, but partly also because we do not want it to be possible), we will, if necessary, render impossible tomorrow (even invoking the will of God as made known in something as vague as 'natural' creation to justify our actions).

Perhaps the most chilling and disturbing example of this is to be found in the film *Blade Runner*[5] where the 'hero' (played by Harrison Ford) is engaged in seeking out and destroying 'replicants' – androids manufactured to live and work on worlds where humans cannot easily live and work, but who escape and come to earth. (This is an earth which has in any case become a nightmare. The film deliberately paints a negative and cynical picture of human beings, whom it compares unfavourably with the androids that it depicts as victims.) What gives the film its power (in stark contrast to the case in *Cherry 2000*) is the fact that the robot replicants are deeply, even profoundly human (only more so). Physically and mentally superior to the human, they are condemned for precisely that, hunted down, and terminated. But the film presupposes (and the audience is persuaded to believe) that these replicants have an inside story, that there is a 'someone' inside them looking out, feeling pain, experiencing mortality, and falling in love.

In such a climate the question, 'Under what circumstances would we be willing to credit artificial life with an inside-out orientation?' becomes even more difficult to answer. Our physical prejudices against the identifiably different (apparent enough in the attitudes of some white Christians to blacks and Jews) sweep all other considerations aside. Identifiable groups 'cannot be human' because they are 'different'; and being 'non-human' we automatically assume that humans have a right to dominion over them.

Somehow we need a way to distinguish an object with an inside-out world from one which lacks such a world-view. To investigate how we might form such an opinion, let us return to the human case and ask what, if anything, leads us to assume that other human beings have inside-out worlds other than the fact that they are recognizably human by birth and that we therefore assume them to be 'like us'.

THE HUMAN CASE
Perhaps the clearest evidence for the inside-out world of a human being can be detected in the grasping of a tiny baby for experience of the world. A baby is like a vacuum-cleaner, sucking everything in, reaching for things, looking, climbing, crawling and walking into trouble at every opportunity as it makes contact with its mother, its cot, its toys, and eventually the wider world. Such a hunger comes from somewhere, manifests the presence of at least an instinct to absorb and grow. Moreover, there is an urgency and an insistence about a baby's learning that draws everything to orbit around it (everything from the intolerable crying that forces mother to get up in the middle of the night and feed it – I would if I could, but I can't so I don't [but I did change the nappy afterwards] – to the terrifying wandering to and fro in front of fires, clambering up unstable cupboards, and sliding up and down life-threatening stairs).

In later and especially adult life this urgency gives way to a more measured capacity to initiate things: processes, conversations, adventures, explorations, flights of fancy. The adult human is a source of initiatory activity. And unlike much animal initiation, which is reactive and based upon immediate need, human initiation is directed towards apparently superfluous and occasionally frivolous ends. We ask about how the world works, why it is as it is, where it came from, and whether there is an ultimate living source of everything worthy of the name 'God'.

The initiatory stimulation that emanates from another human being is perhaps our clearest indication that other humans have inside-out worlds. In another adult human being we encounter an ultimate source of newness, a centre which absorbs (and has throughout its life absorbed) input stimuli from the world (consciously and unconsciously), and has now processed that input data, mixing it into a unique form and combination of words, thoughts and actions, which it now returns to the world in the form of new stimuli which fall upon others.

In encountering another human being we therefore encounter a source which may (and probably will) change us (for good or ill) by bathing us in the outpouring of its-self as manifested in its utterances and actions.

The capacity of certain higher organisms (how high? – who knows?) to act as sources of newness in the world is, in theological terms, a sign of the fulfilment of God's purpose. Something other than God capable of genuine generative activity has emerged in the universe, and one example of that generative activity is to be found in the human species. We seek one another out, albeit selectively, because in one another we find sources of new experience, new feelings, new hopes and dreams. At its deepest level we come to find the centre of our own being in someone else, and we say that we are 'in love'; without the other we are not complete. In some versions of the religious community we are only complete when we live together in large groups, sharing everything, learning to live together regardless of whether we like one another, because in the other, rather than in the self, is to be found the channel for God's grace as God's newness enters into the world.

It is this capacity to make, to start, and to supply that constitutes something approaching the consummation of human being. And because these sources of newness bathe the world in their creations, they do not and cannot ultimately exist for themselves. They must exist for others.

So we approach a definition of divine life (that is, of life as divinely conceived and conceived in terms and in the light of the divine): divine life is the generative source of new creative activity that flows from itself toward the other. To love is to find in oneself and in the other a little part of the image of God.

With this definition of divine life, there is no more reason to doubt or to try to prohibit the possibility that human beings might create sources of genuine newness in their turn than that God would or could do so. If it is a basic instinct to procreate 'after our kind', why not also to create after another kind?

Therefore, despite all the reasons I have given that lead me to doubt whether computer programs as we currently conceive of them are capable of approaching it (that computers do not coordinate convergent impulses; that they are not situated on a long spectrum between full life and death; that they cannot feel genuine pain; and that their programs or neural nets work in ways entirely different from brains) we may reasonably set out one criterion which any such 'life' would have to satisfy in order to be considered life at all: artificial life will have to satisfy our definition of divine life. And in order to do that it will need first to be as hungry for knowledge as a newborn baby in order that, however complex its program, it can be fed with sufficient food to produce new ideas from the depths where perhaps even its programmers cannot understand what is going on.

This is very much the experience of human parents: they watch a vast proportion of the experiences their children have, but those children still amaze them in agreeable and not-so-agreeable ways by producing out of the deep well of their own make-up and stored experiences the first indications of their capacity for generating new ideas and thoughts. The 'programmers' (the parents and the society in which they live) have no idea what enabled their children to produce these things; but produce them they do.

There is a difference here between the pleasure parents enjoy from watching their children do things that are clever (such as playing chess well) and the pleasure they derive from finding that their children are independent centres of creativeness who, even if they still require nurture, are already well beyond the control of any external force (albeit in the grip of the awful fads of adolescence).

An important quibble

It is important that our artificial life satisfy both these criteria: that it be a centre of new creative ideas and that it resource its own ideas by looking for and hungering for new input stimuli. Several programs, especially in geometry, have surprised their programmers by producing novel proofs of results using methods that look as if they are creative, but that is no indication of an inside-out perspective.

An inside-out perspective initiates searches for new information, and the nature of those searches (what it looks for, how it proceeds, what it selects and what it rejects, how it deals with and responds to chance encounters with interesting, relevant and irrelevant material) are indications of the kind of 'person' the inside-out-ness is constituted by.

Science fiction authors, like all novelists, are entitled to invest their characters with inside-out worlds and to claim to know what those worlds are like. Once something has a personal name, imputation of an inside-out world seems almost inevitable. Children do this with their toys. Adults do it in a semi-frivolous way. We once had a car with POD in the number. 'POD' as it became known, quickly acquired a personality for stubbornness, refusal to start and letting us down at inconvenient moments. We told POD's story; it became acceptable to say certain things of 'her' and unacceptable to say others. She had acquired a personality.

We were sad when she was stolen. We were even more sad when she came back. It is not always easy to separate fact from fancy.

Growing minds

Although adults make an enormous investment in the education of their children from the moment they are born, providing them with stimulation to draw out their latent abilities, they rely ultimately upon something over which they are little more than passive observers: the active desire of their child to pick up new knowledge. This thirst for knowledge, whose absence we regard as a symptom of a potentially very serious disability in any child, forces us to watch and wait. We do not program our children the way we program a computer. Computer programming is an active, directed, causal task; this set of instructions will do precisely and exactly this, and barring malfunctions and mistakes there is no doubt about the outcome.

Nothing of the kind is true of children as they grow: not only the stimuli we supply deliberately, but the uncountably vast number of experiences they acquire in their environments without anyone noticing, are absorbed in unique ways which shape their growing minds.

We are close to a major feature of human intelligence: that human learning is self-advancing, self-directing, self-adapting, and autocatalytic in that the more we know the more we tend to want to know. We literally hunger for the knowledge that comes through new and repeated experiences.

Oddly – or perhaps not so oddly – the more we have learned to program or to construct computers with this kind of learning capability, the more we have had to rely upon an increase in passivity on the part of the programmer. Neural nets, for example, work with large number of identical components; their performance involves reinforcing certain connections through repetition; teaching them involves telling them what outputs belong with which inputs and relying upon the nets themselves to 'work out' the outputs for other, different, hitherto unknown inputs. Neural nets are very good at solving what we might call this kind of 'I've been somewhere like this before and I know how to go on' problem. But the nets themselves do not worry about 'knowing how to go on'. They are indifferent to the inputs they are given; even if they are given inputs that were part of their original tutorial they will 'mindlessly' generate the output they were told belonged with them.

We come back to our fundamental question: might a very complex neural net which produced human-like responses to human input (despite the fact that we had absolutely no idea how it did so) ever manage to persuade us that it was 'in there looking out at us'? Or would we insist that it was 'nothing but a machine' and persist with the view that it 'couldn't be anything else'?

The point of this argument is to show up the similarities between the brain as an input-output system we do not begin to understand and a 'sufficiently sophisticated' neural net which, although we made it ourselves, we do not begin to understand either (because it has learnt things in ways we do not understand). As I have repeatedly stated, we treat brains as having an inside-out orientation because we are brains and we have such an orientation, but we have no parallels or analogies (and certainly no personal experience) of what it is like to 'be' a neural net or one of its ultra-sophisticated descendants, and so no basis upon which to form such a judgement.

The question of the soul

Does Commander Data, the android in *Star Trek*, have a soul? Does a human being have a soul? The Sliding Definition Ploy haunts us especially here. What once was called a soul – a thing not made of matter at all, but existing quite independently of the material world, either pre-existent (as in Platonism) or created afresh in each new organism (as some Roman Catholics still teach and believe) – has more recently just become another name for the 'person', that hard-to-grasp 'something' that makes each of us the person we are.

In the latter sense both human beings and Mavis, the domestic robot, have souls; whereas in the former sense (as far as I can see), nobody has a soul. But I do not think that the latter sense is adequate either.

Introspectively, my soul is the seat of the 'I' that has been known to consciousness for as long as there have been conscious creatures. 'Soul' and 'self' are used interchangeably in this sense, and inasmuch as I believe that 'my soul' may survive my death, I mean little more than that something that is identifiably 'me' (to that self and to others) will find itself still in existence after its physical earthly body has ceased to exist.

Selves and consciousness

A human soul is an aspect of the self which exists from-inside-out. It is not simply something you see in me. It is something I see (am aware of) in my self. My soul is a term denoting that totality of somethings which make me what I am in my deepest self, including what I seem to be and the ways in which I should be seen by others and myself. In this sense it seems much less likely that Mavis has a soul, and doubtful that we could tell even if she (it?) did.

Human beings have thought of themselves as made up of three parts – body, mind and spirit (or soul) – for much of human history. In the modern world the body and mind have tended to eclipse the spirit, and in the age of the computer we have tended to assume that the body will in its turn eclipse the mind. Everything will become a random consequence of the brute forces of biology, intricate as they may be.

But a modern equivalent of 'body, mind and spirit' is equally feasible: body, consciousness and self. However interdependent the three may be, they are certainly distinguishable, and what we call 'consciousness' is very close to what we once called 'mind'. Replacement of soul or spirit with self does however cause us some problems. For many of us consciousness and self are the same, and the fact that we equate the two renders us particularly susceptible to the equation of apparently intelligent operations with the existence of selves. We believe in the possibility of computer minds and computer selves because we have come, in our rational way, to think of the rational conscious self as the highest point of evolutionary achievement. But do selves consist in a body's consciousness of itself, or in the deep reservoir of experiences which the body stores that irrupt into consciousness from time to time?

The story I told earlier of my response to my kitchen kettle, the ideas which spring to mind whenever I use it (albeit only dimly to the back of my mind), and the kinds of associations it can so easily set in train, tells the reader far more about my self than it does about my consciousness. And each of us is engaged in just this kind of self-expression all the time. We react to everything in ways governed by a deep reservoir of experiences and associated feelings.

As I sit at my computer typing these words, I am not conscious of the words I shall be typing in a few moments' time; those words nevertheless are latent within my self, even if before they come to be typed they are mixed with extra experiences which are not at present part of that self. The telephone may ring; something on my desk may distract me or remind me of something; a slip of paper stuck to an earlier draft may conjure up a new idea; someone may speak to me; one of my children may come in and ask me something. It is not a matter of saying that the words I am yet to type are already in some sense 'there' in my self, but that my self is a part of the world from which those words will spring. (It is possible, of course, that I shall dry up completely, and write nothing more. We cannot know the future in that respect.)

In that the future does not yet exist, but must be made from components available in the world, the next words that I type are just like new relationships we form, new things we do, new people we meet. We cannot anticipate them, even if we have vague ideas of the range of likely events that may befall us.

When heaven and earth pass away

Do computers provide a way of thinking about the relationship between life and afterlife? Is the abstract nature of a program (rather than the way it is actually put into code in a particular programming language on a particular computer), akin to the existence of a soul in a body?

A full treatment of such a question would require first a detailed analysis of the history of conceptions of the human soul going back into Greek and Hebrew thought, and that is not appropriate in this book. Several things may, however, be said.

First, that the notion of program and hardware seems more amenable to a soul/body dualism than the kind of integrated, embodied notion of the self we currently hold, and that this soul/body dualism is not compatible with the dual aspect theory developed here.

Second, that the kind of system- and implementation-independent program it envisages (a program, that is, which is what it is quite apart from any particular computer or any particular situation in which a version of it is supposed to work) is far more abstract than we might casually suppose.

The first point should by now be clear; the second needs more explanation.

A now-familiar example from everyday life will serve. Let me describe how I heat water in my much-maligned kettle. The sequence of instructions might run something like this:

1 Unplug the kettle.
2 Move to a water supply.
3 Place the spout of the kettle under the tap.
4 Turn on the tap.
5 Fill the kettle.
6 Turn off the tap.
7 Plug the kettle back in.
8 Switch on the electricity.

These instructions are largely – but not completely – system-independent. No specific kettle, water source, tap or electricity supply is mentioned. Many of the instructions, before we could actually perform them, would require additional information (how to unplug, nature of water source, whether there is a tap, how full the kettle should be, where the electricity switch is). Each of these moves us closer to a system-specific implementation.

But how abstract are the eight instructions? What sense would they make, for example, in a world without the concepts 'kettle', 'water', 'tap' and 'electricity'? Clearly none at all. And this is not a mere piece of hair-splitting, for everything we do on this earth, and therefore every program we run on this earth, is in some sense or other earth-specific. That is, every program is an implementation that would make no sense in another cosmos (and in particular in another life lived in heaven) unless we were prepared to insist that the same kinds of ends (such as boiling water) would remain requirements.

The point, quite apart from water-boiling, is that the kind of completely system-independent abstraction that would translate readily into another life in heaven in such a way that it was indifferent to the changed circumstances of the 'beings' existing there, is extraordinarily vague and difficult to grasp.

What qualities might be capable of surviving death by virtue of their system-independence? It is easier to give a list of things that could not. Most kinds of human expertise are emptied of meaning once their earth-implementation is forgotten. This applies to science (because there is a new heaven and a new earth); economics (because there will be neither rich nor poor, slave nor free); linguistics (for the separation occasioned by language will cease); sporting prowess (for will the spiritual body play cricket?); engineering (for what materials will be as we know them now, and what buildings will they be used to build?). The list is a catalogue of things we will not need in the next world. And even if mathematics survives as a set of abstract concepts like number, what will we be concerned to measure or to count?

If we are to seek the kingdom of heaven and the things that are of God, what are those things? Will these three abide: faith, hope and love? Will we need faith when God is before our eyes? Will we need hope when all things are accomplished? Will we need love when all things are made one by being reconciled to God?

Is then the last thing that can survive our death our very self?

What needs to be understood here is that what Joseph Weizenbaum once said of artificial intelligence – that since all intelligence is context-specific and relative we should not be looking for a generic artificial intelligence but only an ability to perform certain tasks in certain contexts – is also true of the notion of the self.

Like the water-boiling example as extended to the earth-specific nature of all programs, the notion of a self in the absence of an earth in which to make its-self known scarcely makes much sense. Selves, like intelligence, are context-specific. How much of what someone is would remain intelligible in the absence of the kinds of concerns that are all around us in our daily lives: job, money, houses, possessions, skills, knowledge, bodily appearance and so forth? Can even the self survive the dissolution of the earth and its own body as its fundamental hardware? Is even the self that 'abstract'?

Part of the difficulty we have is that we believe that we know ourselves, but we do not. I am to myself rather as a table is to me, as part of my world. Aspects of my self keep popping into view and then I lose sight of them. I have memories of the kind of person I am, but I am not permanently totally 'to hand' (in full view). The self is present to consciousness as facets of a complex object.

To try to keep hold of our sense of self we tell stories which weave the self to which we have given a name into a narrative involving other persons and objects and events in the world.

But what sort of a story could we tell that was independent of those other objects and events, even if (in heaven) we were able to incorporate other persons? What kinds of interactions with other persons could we possibly engage in that might form parts of our stories?

I think it is clear that whatever problems the software/hardware distinction is intended to solve are not really solved because the software is not system-independent, and cannot be, once we allow for the dissolution of the earth. Whatever the new heaven and the new earth may hold, they will require new skills and concepts that may need to be learned afresh, in which case the old programs will be of little use.

The system-dependence of software involves far more than the fact, for example, that software that will run on one computer will not in general run on a computer with a different design. The 'system' includes the complex mesh of problems the software and hardware together are used to solve. Quite specifically, most of the things human beings learn, and all the things that they acquire, on earth, are so tied to the earth that they would count for nothing in a completely different world such as heaven.

The death of Jesus, naked and alone, with nothing that the world counts valuable to 'comfort' him, points directly to the illusory nature of much that we take to be permanent. Most of our learning loses all significance once stripped from the earthly context. The same is true of our wealth and power. Such candidates for the afterlife as we can imagine are more closely related to the background associations of actions – love, loyalty, honesty, justice, faith and hope – than with the things those actions employ. What I think of my kettle is more significant in these terms and in this context than the apparently more solid and permanent kettle itself. Yet those associations are precisely the things which software in the computer sense is least able to generate.

We seem in real danger of having to acknowledge that, quite apart from any problems associated with the death of the body as that in which each self is embodied, the concept of the dissolution of the earth, in relation to which each self reveals and defines its-self, is an even greater challenge. It appears that the self dissolves into the infinite spaces created by the voiding of all the contexts within which it can be shown.

Can Commander Data go to heaven?

Our religious (or secular) prejudices might lead us to answer this with a firm 'no!' We might similarly (although a good deal less firmly) give the same answer to a question such as 'Do dogs go to heaven?' (although I can think of several people for whom the opposite answer would be given even more firmly).

I would want to turn the question round, and with it to alter the entire universe of discourse in which we consider such issues.

Let us suppose that an android such as Commander Data from *Star Trek* (but not, perhaps, the android Cherry 2000) spends a considerable amount of time with us (not necessarily on a starship). The sophistication of its responses to us and its memories of us would come to be repositories of significant portions of those experiences and qualities which constitute 'me'. And if I am to be my full self in the afterlife, I will need to have access to those aspects of my self which are only accessible (because only reconstituted in full) when I relate to 'living' Commander Data. The question, therefore, is not 'Will Commander Data go to heaven?' but 'If Commander Data does not go to heaven, how much of the self which resides in my relationship with Commander Data will not go to heaven either?' (And that presumably supplies an answer to the question of the dogs too.)

To the extent that who I am is wrapped up in my relationship with who you are, and therefore resides not so much 'inside me' or 'inside you' as in what happens – what is made – when we are together, to that extent if you do not go to heaven, that aspect of my self cannot go to heaven either. So 'Never send to know for whom the bell tolls; it tolls for thee' may be more true than even John Donne could have supposed.

If, therefore, a time should come when significant portions of human life are devoted to relationships with androids, we may find ourselves less concerned to argue that such creatures should not or cannot enter the gates of heaven, than to plead with St Peter that if we are to be admitted, they must be admitted too.

There is no doubt that the universe has brought forth creatures with minds. These creatures have been able to lift those minds to a level where they have presumed to be capable of knowing and naming God. In the person of Jesus Christ they have come to believe that one particular inside-looking-out-ness was not merely that of a human being, but that of God. The one who saw the world with Jesus' eyes was God; the one who suffered with Jesus' body was God. And this was not a mere 'seeming', but a real and genuine suffering such as we can each suffer. Because of Jesus, God knows what it is like to be a human being as a human being, from-inside-out, rather than as God, from-outside-in. The only God Jesus knows is the God Jesus is.

It may be that the only way God could create independent centres of agency to stand over against himself who were genuinely free, was to create them in a world in which there would be gradual, painstaking, fragile and fallible evolution. In that case, God might be said to have grown himself children rather than to have created them.

Those children were not made to be alone. In knowing one another they come to know themselves, and they perhaps can know themselves most fully when they know themselves in God. For then they know even as they are known.

And as these creatures grow into one another more and more, and so come to define who they are in terms of their relationships with others and with their God, it may be that we find their selves redefined in terms of those relationships. In that case, it may be that it is what is known of us by others and by God that supplies the repository from which our recreation in the new heaven and the new earth may arise.

6 – THE UNIVERSE, EVERYTHING AND LIFE

In the beginning

In the beginning, God created the heavens and the earth. And he had a problem. In fact, he had several.

Out of the deep wellspring of his love, he decided to create life which was not himself. This life was to be made in his own image. It would be intelligent, creative, capable of independent thought, aware that it is alive, and free. The first problem is: how can this be done? How can something be made which is not God but which shares so many of God's qualities?

At first, it may seem that all that needs to be done is to make such a creature and to give it some of these properties. But that is the problem: God knows how to make such creatures; they are what human beings have thought of as angels in some traditions. But they are not free; they are 'yes-persons'. They are wonderful in their way (particularly the glint of gold in their outstretched wings), but they are not quite what he wants to make. Angels love him because they cannot do anything else. Angels are, so to speak, programmed to love God. Their love is, in a way, God loving himself. God wants something more: the free response of independent creatures to the experience of life. Programming them to love him just will not do.

But even before the first problem has been solved, God becomes aware of another one. Unlike the angels, who can be relied upon to behave (most of the time), the kinds of creatures he has in mind will, simply by virtue of the very qualities he wants to give them, be able to refuse to behave. The artificial life-forms may turn out to be monsters. They may ill-treat one another. They may even learn to destroy one another. They may refuse to love him. They may even refuse to believe that he exists.

And suddenly yet another problem arises: that if they knew that God made them, he could never be sure that their love was freely given. They might simply pretend to love him, fearing the consequences if they did not. So this artificial life not only cannot be programmed to be free; it must also live in ignorance of the existence of its creator. At least, it must not be able to rely upon indubitable evidence of the existence and nature of its creator. It needs to come to believe in and to love God by the direct operation of its own free independent mind. This creature must only be able to find its full essence by becoming a creature that believes in things that are uncertain. It must become a creature of faith.

God has these three problems: how to create artificial, intelligent, independent, aware and free life; how to ensure that such life will not prove to be a monster; and how to hide his own existence from it sufficiently to ensure that it believes in and loves him for the right reasons.

The first two of these problems are almost exactly equivalent to the problems faced by human attempts to create our own kinds of artificial, intelligent life. We are not sure whether it is possible; we are unsure how, even if it possible, we can achieve it; and we are more than a little concerned that if we succeed, that life may prove to be monstrous rather than benevolent.

God solved all three problems in ways that are as far removed from the idea of a computer program as one could imagine. He solved them by creating the universe, everything, and life. He did not program life; he grew it. He did not manufacture it; he evolved it. And all the fears he had for it and about it proved all too well-founded. Many of the creatures that ensued turned out to be veritable monsters. And the most monstrous of all (as far as we know, which is not, actually, very far) turned out to be humankind. Yet God had solved the third problem by hiding himself, and that seemed to work rather well; men and women of faith did evolve, did come to imagine his existence and to believe in him. Not many perhaps, but enough to make the whole project seem worthwhile. The chance of life had proved to be better than no chance at all.

But the monsters – and all the creatures were monstrous in some, albeit minor, respects – meant that God had a fourth, and final problem: how to draw his children back to the true path without revealing himself so unambiguously as to make the third problem worse. How, having created artificial life, could it be called back from self-destruction?

God's response to this problem was brilliant; a stroke of boundess genius. He chose to reveal himself to his creatures in a way that would leave the ambiguity in place, that would introduce into the world a new stream of thought originating directly from himself, without the use of intermediaries (lawgivers, prophets, priests and miracle-workers). God chose to become a human being. He who was the only true life entered into the artificial life he had created, taking for himself a body such as his artificial life enjoyed. And in doing so he also solved another problem: that while as God he could know whether his creation had achieved the kind of awareness he had intended, he could not feel what that awareness was like. As God he was shut out from the feelings of the artificial life he had created, their pains, their sorrows, their joys. But in becoming that artificial life, he acquired the same orientation to the world that each of his creatures enjoyed. And so by becoming human, God discovered what it felt like to be a creature, and in doing so he took upon himself in the only way that he could the means to answer the question: can these bones live?

Can, in other words, this artificial creature ever reasonably be said to have an inside world, an awareness? And is the expectation God has laid upon his shoulders, that this creature should, solely by living in the universe, come to know of, and to believe in, and to love the God who made him, is this expectation reasonable?

Similarly, the only way a human being who seeks to create her own form of artificial life and intelligence could definitively answer the question, 'Is this creation aware?' would be to become that creature. This is what God did. This is how God solved and answered perhaps the most profound of all questions: can these bones live? 'The Word became flesh.'

It seems that human beings are retracing the steps of this process. We are gradually realizing that our early attempts to manufacture and to program artificial intelligence and life were moderately successful but ham-fisted ways to solve the problem God solved differently. And we are now realizing that, if we are to solve the same problem, we shall have to grow and evolve our artificial life-forms, rather as God did.

We do not fear that our children will turn into quite the same kinds of monsters as this artificial life may produce. Within (admittedly rather broad) limits, we know what sorts of creatures our children will be. They involve reproduction 'after our own kind'. But artificial life will be nothing like our biological children. (At least, it will be a profound irony if it turns out that the only way to produce artificial life is to produce biological children.) With artificial life and intelligence we are trying to solve a problem similar to that which God first contemplated: how to create an intelligent, innovative, aware, free creature capable of independent thought.

But perhaps this is not our intention. Perhaps this is to read our attempts in altogether too benevolent a fashion. Perhaps all we really want to do is to create a robot that will relieve us of all the things we think tedious, dangerous, laborious, and so forth. Perhaps God's problem was very different, in that he envisaged that his artificial life would be capable of love where we are interested only that it be capable of work. Perhaps the deepest question we should ask ourselves is not whether we should, or how we can, or to what extent we might be able to, create artificial life, but whether we are yet mature enough as a species to shoulder the responsibilities that such an action will involve.

It is tempting to reply to those who express doubts about artificial intelligence by saying, 'Of course there is such a thing as artificial intelligence; we have been using it for thousands of years.' This is not altogether frivolous. Our intelligence, at least as measured relative to God's intelligence, is highly artificial. Faced with a pile of carbon atoms, 'bones' as we might call them, the question, 'can these carbon atoms be intelligent?' seems to answer itself: 'Don't be ridiculous!'

The situation is not so very different with strings of noughts and ones in a computer, or with electrical and chemical discharges in neurones. 'Can these bits think?' and 'Can these neurones think?' and 'Can these bones live?' are not fundamentally different questions. They boil down to 'Can that which is not God acquire some of the characteristics of God?' And the Bible and the Judaeo-Christian tradition accept some version of the answer given in Genesis: God made man 'in his own image' from the dust of the ground.

So, contrary to what Christians usually suppose, far from it being blasphemous to think that we might make artificial intelligence or artificial life, we have a precedent that could not be more authoritative: God made us in his image from the dust of the ground. Modern science alters the detail with a long story about the Big Bang, the emergence of stars and heavy elements, and the eventual evolution of life, but it does not deny that, in the end, we are made 'out of the dust'. Neither does it deny that we shall return to the dust. Nothing is more certain.

Having created his creatures and called them back from almost certain self-destruction, God's problems are not over. The fragility of the bodies his creatures have makes their death inevitable. In fact, death turned out to play an essential part in the evolution of life. But the inevitability of death seems to render all that God has achieved rather pointless. A brief life in which the clay comes to know God is made empty if it still returns to be merely clay. So God's final problem is this: how, having made and redeemed intelligent, independent, aware creatures which exhibit freedom, he can rescue them from the inevitability of death.

The problem is made more difficult because the minds which those creatures have grown are inseparable from their brains, and their brains are things made of the dust. When the brain dies, the mind dies. The death of the body implies the death of the mind, the spirit, the soul, of every human being, of every creature. Yet it offends God that something so precious, something that represents the crowning glory of his creative work, should suffer the ignominy of oblivion. The loss of such creatures pains God.

The solution arises in the difference between making something that is first known and knowing something that is first made.

MAKING SOMETHING THAT IS FIRST KNOWN

To make something that is first known (known beforehand), we must first have a clear plan, and know how to effect what we are making. Such a making, as the potter moulds the pot, is very direct. What is made lies under the immediate control of the maker. The knowing must come first, even if the exact shape – the exact pot – is altered slightly as we proceed.

To make something that is first known may nevertheless give rise to surprises. A pot may come to assume an artistic significance its potter did not envisage. A typewriter or word processor, despite the precise way in which it is made according to a previously known specification, will nevertheless be used to produce books and diagrams its designers could not have anticipated. But the fact that an artefact can surprise its makers or be used in surprising ways is not proof that it has a life of its own, only that they were unable to anticipate (project) all the ways in which it might be used or all the things it might do. And whatever it did could always be explained (retrojected) in terms of unrecognized potential in its original design.

What if the challenge were different? What if the challenge were to make something that could behave in ways so radically different from those at first envisaged that it could be considered to have a life of its own? In those circumstances we would need first to make it and then to know what it was that we had made.

KNOWING SOMETHING THAT IS FIRST MADE

To know something that is first made involves a much more difficult process, for how can something be made before it is known? It is this kind of making that is creative; from it arise things which are new, which were not known before they were made. This kind of knowing involves allowing some process to take its course, free from direct interference, and being ready to know the result for what it is, for what it evolves to be. This is the making that gives rise to independent centres of free thought, to minds.

The difference, however, is deeper than this. If that which is made is to be free, it cannot be known in advance, for that would be to prescribe what it was to be. Instead, having allowed it to come into being, God can then know it for what it is in its freedom. And knowing what it is in its freedom allows the same thing to be remade in the same freedom. Because God's creatures have no thoughts that are not brain thoughts, and no minds that are not brain minds, every creature's mind can be reconstructed from the configuration of that creature's brain. God need not interfere; he need merely know in order to recreate faithfully the mind of the creature which God's universe has evolved and made.

The resurrection

In other words, it is because there is only one kind of stuff, and that a kind of stuff that produces minds, that minds can be recreated after the death of the body. The 'body' in which they are recreated is called, ineffectually, 'the resurrection body'. It is not possible to say in what that body will consist.

How does such a brain differ from the brain which God could have made for himself? It differs precisely insofar as the state of that brain has itself been changed according to the wishes of the mind which depends upon it. This is the feedback process by which each creature defines its own being. This brain is not a brain designed by someone else (however much the wiring may be configured by something else); this brain is what the person whose brain it is has made it.

And if you ask whether perhaps all this was necessary, whether, that is, God could not simply have short-circuited the whole process by manufacturing every conceivable brain configuration, I would give two (of many) answers. First, that God willed it otherwise, and therefore willed that his creatures should be self-determining, particular, and unique. God did not choose to make every kind of creature; he chose to make and to have evolve only some. (That in itself says something important about God.) Second, it depends upon whether we are prepared to allow the notion of complexity to reach even as far as God's foreknowledge. It may be, that is, that even God could not anticipate the actual details of the level of complexity that would be necessary to generate free, independent, aware sources of creative life. In that case, for God simply to have 'thrown together a few zillion neural connections' may have been as unlikely to produce such a creature as a human being. This is certainly true for us. We know that monkeys sitting at typewriters will not produce *Hamlet* in the life of the universe. The complexity of the brain may mean that even God could not produce Shakespeare in all eternity. It all depends upon what we are prepared to believe of God. The fact that God did not snap his fingers and make us all in an instant may as easily imply that he could not do so as that he chose not to do so. Only God knows the answer, but the evidence points to my answer: God did what he could. Let us leave it at that.

The analogy is with the programmer of the neural net or the genetic algorithm. Neither knows exactly what she is looking for, but when she has a solution she can reproduce it. The training of the neural net leads to a solution to the problem in hand which could not have been programmed in traditional ways; once we have that solution, however, we can reproduce it easily by simply copying the network and the weights that have evolved in each connection. The *analogy* (I do not feel it to be more than that) helps us to understand both why it may have been necessary for God to grow minds, and how he might save them from oblivion having done so. It also indicates that the analogy with the more traditional program running in hardware, based upon specific instructions, is *not* a helpful way to think of the relationship between brains and minds.

All this seems to point to only one possible answer to the tongue-in-cheek question, 'Can Commander Data go to heaven?' Since Commander Data will be made, like us, out of the dust (albeit dust we have shaped rather than God's own evolutionary process), there is no reason in principle why Commander Data should not be reconstructed in heaven. Only God can probably judge whether he has an inside-out, but we too will have a view based upon the 'It takes one to know one' principle. And we may also find that our own mental existence is so tied up with his that not to have him around in heaven will seriously diminish our enjoyment of being there (as much as missing our favourite dog or cat, our husbands, wives, children, and dearest relatives and friends).

Because I love you, I cannot be without you. Where I am to be, you must be also.

Epilogue

I suspect that this book will have satisfied neither the sceptics nor the enthusiasts. The anxious will not have heard me say, 'No, it will never happen.' The optimists will not have heard me say, 'Yes, it will certainly happen.' Perhaps such disappointment is the fate of any attempt to be fair to both sides of a complex argument.

Is it possible to come to any conclusions?

However many its setbacks, artificial intelligence will progress beyond any limit we currently think it reasonable to set. It will do so largely because it challenges us. The more difficult it seems, the more determined we shall be to conquer it. Whether it will ever manage to create an artificial inside-out world that will exhibit (and, more to the point, enjoy) all the properties of mind, I rather doubt. But the tantalizing question, 'How would we tell?' leads me to a final, equally tantalizing thought, with perhaps more religious significance than anything else I have said.

If an android were one day to present us with evidence that it had a mind, it is inconceivable that that evidence would be irrefutable, indubitable, and therefore irresistably convincing; sufficient, that is, to convince even the most determined sceptic.

The question of the possibility of such minds is therefore, I think, rather like the question of the possibility of the existence of God. No evidence can convince the absolutely determined sceptic. But fortunately, neither the question of the existence of God, nor the possibility of artificial minds, depends upon our willingness to believe in it.

The success of artificial intelligence in one form or another will exceed most current estimates. Machines of some description will one day, although not for some considerable time, be capable of passing a Turing Test at least to the extent that they will make more entertaining and intelligent companions than many human beings. A vast number of skills which we currently believe to require human ingenuity and insight will succumb to automation, squeezing human expertise into narrow fields where automation is either too difficult or too expensive to contemplate. This has already happened in the machine-tool industry and mechanical engineering; it will eventually be true of all human life and work.

Nevertheless, all these advances (if such they be) will amount to nothing more significant in terms of the souls of machines than the evolution of transportation from chariot to spaceship They will not provide the machines on which they work with inside-out worlds in any sense at all. When Gary Kasparov lost the first game of his six-game match against IBM's Deep Blue early in 1996 – the first time a reigning world chess champion had lost against a computer under tournament conditions – he wrote that he had sensed 'a new kind of intelligence'. Quite possibly, he did, but this was not an intelligence of which Deep Blue itself was aware in any sense at all. Deep Blue has no inside-out world, no consciousness, no mind.

This may reassure us. I am not convinced that it should. There is at least a hope that, as with human beings, a machine with an inside-out, a world-orientation, a mind, might be susceptible to reasoned appeals to something akin to its better nature. It might, eventually, develop a social conscience. We cannot appeal to the 'better nature' of a machine without an inside-out world: it does not have one. Awesomely intelligent computers, so complex that nobody fully understands their workings, and which have no minds, would arguably be more terrible than those that did have minds, should they ever prove possible. Yet such machines are already very nearly here, controlling international stock, bond and currency markets: machines mindlessly and stupidly applying their programs to matters influencing life and death. There is, in the short term at least, more reason to worry about them than about machines with artificial intelligence.

And what of the prospects for machines with minds, consciousness and souls? I will venture a conjecture. Although they will make a significant contribution to the peripheral requirements of such developments, symbol-processing computers running programs written in languages such as C, Pascal, Lisp or Prolog will never generate minds, however intelligent their performances become, however well they convince us that they have minds, and however willing we are to be fooled.

Instead, to be successful, the AI community will need increasingly to follow God's approach to creation: to set up situations in which minds, rather than being programmed, can grow and be grown, where their growth cannot altogether be understood or its future predicted, and where the structures of the machines in which they grow become closer and closer to those of the organisms which created them.

As my final chapter argues, having decided to make a universe other than himself, and creatures with intelligence and freedom, God was faced with several deep problems – problems similar in many respects to those that confront the AI community. In particular, he had to solve the problem of how to engender an inside-out world in his creatures while leaving them the freedom and therefore the creativity and originality to be worthy to stand over against him, less as servants than as friends.

If images of the Tower of Babel, Prometheus and Frankenstein temper my excitement and enthusiasm for the cause with caution, I take no comfort from the false starts the AI community suffered, and I think that those who laugh at its early fumblings and exaggerations will laugh neither last nor best. We have already passed the lowest point of AI's fortunes, and significant breakthroughs will come. The complacency of those who think the task impossible, or its implications and consequences of no account, worries me far more than the thought of countless brilliant and ingenious scientists chipping away at the problems year by year. If there is reason to fear, it is always later than we think.

Despite the precedent set by God's creativity, I am sure that some religious people will choose to convince themselves for very sound, spiritual reasons, that genuine artificial life with an inside-out world cannot be achieved. They will insist that there will never be a mind machine. But the apparent impossibility (as it may seem to us) of artificial minds did not deter God from making them. For the universe is God's Mind Machine.

Notes

1 Joseph Weizenbaum, *Computer Power and Human Reason*, W.H. Freeman, 1976, p. 277.

INTRODUCTION
1 Cf. my 'Faith's Third Age' in *Colloquium*, Vol. 27, No. 2, Nov. 1995, pp. 109-128 for a longer discussion of the analogies between childhood, adolescence and adulthood and our attitudes to science and belief.

CHAPTER 1
1 The opening chapter of Jack Copeland's *Artificial Intelligence* (Blackwell, 1993) contains a readable and authoritative summary of both the achievements and the exaggerations.
2 Roger Penrose, *The Emperor's New Mind*, OUP, 1989; Roger Penrose, *Shadows of the Mind*, OUP, 1994.
3 Cf. my 'Information Theory, Biology and Christology', in Mark Richardson and Wesley Wildman (eds.), *Building Bridges between Science and Theology*, Notre Dame University Press, 1996.
4 Thomas Nagel, *The View from Nowhere*, OUP, 1986.
5 Cf. Copeland, 1993 for an excellent discussion of these issues.
6 See Roger Sperry, *Science and Moral Priority*, Blackwell, 1983; John Eccles and Karl Popper, *The Self and Its Brain*, Springer Verlag, 1977; and Steven Rose, *The Conscious Brain*, Penguin, 1973 for details.
7 Maureen Caudill, *In Our Own Image: Building an Artificial Person*, OUP, 1992, p. 191.
8 Ludwig Wittgenstein, *Zettel*, edited by G. E. M. Anscombe and G. H. von Wright, Blackwell, 1967.
9 Gilbert Ryle, *The Concept of Mind*, Penguin, 1949, p. 188.

CHAPTER 2

1 Benedict de Spinoza, *Ethics*, George Bell and Sons, 1906, Part II, Para VII, Note; Stuart Hampshire, *Spinoza*, Pelican, 1978, p. 63ff.; Max Planck, 'Phantom Problems in Science', reprinted in Dagobert D. Runes, *Treasury of World Science*, Owen, 1962, p. 848ff.; Roger Collingwood, *The Idea of History*, OUP, 1992, p. 213; but cf. Thomas F. Torrance's criticisms of Collingwood in *Theological Science*, OUP, 1969, p. 316ff. ; Michael Polanyi, *Knowing and Being*, Routledge and Kegan Paul, 1969, p. 220; Teilhard de Chardin, *The Phenomenon of Man*, Collins, 1965, p. 60; Donald Mackay, *Brains, Machines and Persons*, Collins, 1980, p.16f.; and Thomas Nagel, *The View from Nowhere*, OUP, 1986.

2 Quoted in Daniel C. Dennett, *Consciousness Explained*, Penguin, 1991, p. 412.

CHAPTER 3

1 Frank J. Tipler, *The Physics of Immortality*, Macmillan, 1994.
2 Richard Dawkins, *The Selfish Gene*, Paladin, 1978.
3 See M. Mitchell Waldrop, *Complexity* (Penguin, 1992) and William Poundstone, *The Recursive Universe* (OUP, 1987) for more details.

CHAPTER 4

1 See Stephen Levy's *Artificial Life* (Jonathan Cape, 1993) for details.

CHAPTER 5

1 Douglas Adams, *The Hitch-Hiker's Guide to the Galaxy*, Pan Books, 1979.
2 Roger Sperry, *Science and Moral Priority*, Blackwell, 1983.
3 *Cherry 2000*, released in 1988, directed by Steve DeJarnatt, and starring Melanie Griffith and Harry Carey.
4 Isaac Asimov, *I, Robot*, Dobson, 1967.
5 *Blade Runner*, released in 1982, directed by Ridley Scott, and starring Harrison Ford, Rutger Hauer and Sean Young.

Further Reading

Bryan Appleyard, *Understanding the Present*, Picador, 1992.
Ian G. Barbour, *Religion in an Age of Science*, SCM Press, 1990.
Margaret A. Boden, *Artificial Intelligence and Natural Man*, Harvester, 1977.
Margaret A. Boden (ed.), *The Philosophy of Artificial Intelligence*, OUP, 1990.
Patricia Smith Churchland, *Neurophilosophy*, MIT Press, 1989.
George V. Coyne and Karl Schmitz-Moorman (eds.), *Origins, Time and Complexity*, Labor et Fides, Geneva, 1993.
Willem Drees, *Religion, Science and Naturalism*, CUP, 1996.
Hubert L. Dreyfus, *What Computers (Still) Can't Do*, MIT, 1992.
Gerald Edelman, *Bright Air, Brilliant Fire*, Penguin, 1992.
Anthony Flew (ed.), *Body, Mind, and Death*, Macmillan, New York, 1964.
Mary Midgley, *Wisdom, Information and Wonder*, Routledge, 1989.
Mary Midgley, *Science as Salvation*, Routledge, 1992.
Hans Moravec, *Mind Children*, Harvard University Press, 1988.
Seymour Papert, *Mindstorms*, Harvester, 1982.
Derek Parfit, *Reasons and Persons*, OUP, 1984.
Arthur Peacocke, *Theology for a Scientific Age*, SCM, 1990.
Michael Polanyi, *Personal Knowledge*, Routledge and Kegan Paul, 1958.
John C. Polkinghorne, *Science and Providence*, SPCK, 1989.
John C. Puddefoot, *Logic and Affirmation*, Scottish Academic Press, 1987.
Richard Rorty, *Philosophy and the Mirror of Nature*, Princeton University Press, 1980.
Andy F. Sanders, *Michael Polanyi's Post-Critical Epistemology*, Rodopi, Amsterdam, 1988.
John Searle, *Minds, Brains and Science*, BBC, 1984.

Thomas F. Torrance (ed.), *Belief in Science and in Christian Life*, Handsel, 1980.

Christoph Wassermann, Richard Kirby and Bernard Rordorff (eds.), *The Science and Theology of Information*, Labor et Fides, Geneva, 1992.

INFORMATION ON THE INTERNET

There are many internet sites related to AI. For general pointers to other resouces, try:

http://ai.iit.nrc.ca/

For some indications of the military interest in AI, try:

http://www.aic.nrl.navy.mil/

For many resources, and access to some technical papers on artificial life, genetic algorithms and so forth, try:

http://www.santafe.edu

Index

Adams, Douglas 95
Afterlife 118–23
Agency (self-direction) 61–5, 95–6, 123
Algorithm 65–6, 72
Android 4, 6, 55, 74, 75, 100–1, 108, 115, 122, 133
Angels 124
Animals, consciousness of 46–7, 100–1
Artificial intelligence (AI) 5
  challenge of 17–19
  divine creation and 125–7
  expert systems 73
  future of 8–12
  genetic algorithms 72
  history of 7
  human intelligence and 124–30
  neural nets 71–2
  practical uses of 7–9, 73–5
  problems of defining 3–5, 57–67
  strong and weak 31, 59, 70–1
  see also Intelligence; Frame problem
Asimov, Isaac 108
Awareness 10, 26, 31, 42, 46, 48, 55, 79, 126

Babel, Tower of 17
Background assumptions 24–6, 121
Background processing 37, 53, 79
Bannister, Sir Roger 22
Barbour, Ian 84
Bible, the 128
Biology 5, 68, 75, 83–4, 91, 116

Blade Runner 109–10
Brain 39–56
  collisions, correlations and 32–8
  compared with computer 31–8
  feedback and 47–50
  generates mind 98
  influenced by mind 98
  inside/outside distinction 42–5
  see also Mind–body problem

Caudill, Maureen 32, 51
Chaos below thought, the 36–7
Cherry 2000 106–9, 122
Chess 14, 63, 66, 75, 76–8, 80, 87, 134
Chinese Room argument, *see* Searle, John
Christianity
  compatibility with science 81–3, 89, 91–4
  pre-existence of soul and 41
  significance of incarnation 103, 121
Clarke, Arthur C. 11
Cockroach (and taste of scrambled egg) 27, 103
Collingwood, Roger 43
Collisions 32–5
Communication, computers and 8–10
Companionship, machine and human 105–9
Complexity 8, 32, 59, 64, 96, 131
  theory of 72

Computers
  ability to read and write 53–5
  chess-playing 76–8
  compared with growth of children 114–15
  limits of hardware 23
  machine life 103
  minds 99–100
  software 16–17, 22–3, 76–8, 121
  symbol processing 72
  thought and 30–31
  Turing machines 12–15
Consciousness 6, 14, 35, 42–55, 70, 74, 79, 96, 116–17, 120, 134–5
  and computers 49–50
  defining 46–50
  recognition of in others 12–14, 50–1
  uniqueness of 36–8, 103
Conway, John 62
Copernicus, Nicholas 84
Correlations 32–3, 38, 48
Creativity 50–3, 66–7, 88, 135–6

Darwin, Charles 41, 85, 86
Data, Lt Commander 3, 6, 11, 115, 122, 132
Dawkins, Richard 57, 62, 82
Death 46, 76, 83–5, 87, 91–4, 112, 116, 119–21, 128–30, 135
  and resurrection 132
Descartes, René 31
Disease, theological 81–4
Doctor Who 11
Donne, John 122
Dreyfus, Hubert 32
Dual aspect theory 42–5, 48, 101, 118
Dualism 39, 118

Eccles, Sir John 22, 86
Einstein, Albert 82
Emulation 65–6
Evocation 44
Evolution 5, 15, 67–8, 72, 81, 83, 86, 89, 93, 123, 128, 134
  God and 125–9
  of artificial intelligence 127–9
  retrojection 68
Expert systems 73
Explanation 47, 64, 82, 88, 118

Faith 83, 120–1, 125
Feedback 46–53, 130
Finitude, awareness of 92
Frame problem 23–7, 29, 73, 108
Freedom 41, 97–9, 128, 130, 135

Galilei, Galileo 84–6
Genetic algorithms 5, 72, 131
God
  artificial life and 79, 112
  become man 126
  creation by 95, 124–9
  evolution of life by 131–2
  knowing from inside out 101–3
  playing dice 82
Greek mythology 17

HAL 11
Heaven 6, 40, 41, 118–23, 132
  see also Afterlife
Hebrew notion of mind and body 40–1, 44
Hofstadter, Douglas 49
Human beings
  hardware/software analogy 76–8
  as machines 60–1
  mind, body and spirit 116–17
  mystery and 79–81

pain and suffering of 91–4
purpose and 78, 95–6
Hume, David 46

IBM 8, 134
Ignorance 79–81, 87–8, 90–1, 125
see also Mystery
Information 12, 23, 48, 57, 62, 73, 106, 113, 119
theory of 83–4
Inside-looking-out-ness 12, 18, 25, 32, 63, 66, 75, 79, 100–2, 123, 133–6
and creativity 50–5
defined 42–5
hard-wired 15, 32
and inside story 108–16
and one-kind-of-stuff views 96–9
and *qualia* 27
and thoughts 30, 34
Intelligence 19, 63–70
and agency 64–5
explanation and prediction and 67–70
machines and 80–1
simulation and emulation of 65–7
see also Artificial intelligence; Turing test
Internet, the 9, 83, 84
web sites on AI 140
Interpretation 13, 54
Intuition 65, 77, 80

Jesus Christ 102
death of 121
inside-looking-out of 123
resurrection 40–1

Kasparov, Gary 77, 80, 134
Kettle, and Frame problem 25–7, 117–21
Knowledge 14–16, 20, 22, 24, 27, 32, 67, 73, 82, 101–2, 108, 112–14, 120
embodied and learned 15–17, 32
mystery and ignorance and 79–81
of other minds 100-1
Kubrick, Stanley L. 11

Language 15, 40, 54, 65, 73, 75, 93, 96, 100
and Frame problem 23–7
constrains the way we think 21–6
earth-bound 118–19
and learned knowledge 16–17
of lost civilization 29–30
and Sliding Definition Ploy 57–8, 61–2, 115
Life, artificial 3, 50, 62, 78, 92, 100–1, 136
human life as 124–8
and morality 109–13
Life, purpose of 95
Love 120
machine companions and 106–7

Machine 3–4, 8, 11, 19, 23, 26, 30, 43, 45, 51, 55, 64, 73, 82, 115, 134
God's mind machine 136
knowledge of life 103–6
as pejorative term 77–80
problem of definition 57–61
simulation and emulation in 65–7
Turing machine 13–17
Mackay, Donald 43
Mavis (domestic robot) 74–5, 116
Meaning of words, the problem of 3, 13, 28, 29, 30, 53, 54, 57–8, 95, 119
Memory 16, 46, 51

Mind/body problem 39–40
  dual aspect theory 42–5
  growing minds 114–15
  one-kind-of-stuff 96–7
  two kinds-of-stuff 39, 96
  where is mind? 97–8
Models 12, 32, 36, 66, 72, 107
  inversion 19–21
Mystery 2, 46–9, 79–81, 87

Nagel, Thomas 27, 43
Neural net 5, 8, 31–2, 51, 71–2, 99, 103, 112, 114–15, 131
Neurological activity 36–7
Neurones 32, 72, 96–7, 128
Neurophilosophy 43
Nurture 113

Orientation, mind as 37, 41, 43, 50–1, 109–10, 115, 126, 134
Outside-looking-in-ness 44, 49, 55, 63, 101–2, 123

Pain, problem of 70, 91–3, 100, 103, 112
Parallel Distributed Processing (PDP) 72
Peacocke, Arthur 84
Physics 5, 8, 28, 57, 75, 83–4, 91
Planck, Max 43
Plato, platonism 39, 41, 115
Poetry, and machine creativity 14, 53–5, 67
Polanyi, Michael 43
Polkinghorne, John 84
Prediction 82
Prejudice, religious 78, 82, 88

Qualia 27–9, 38, 47–9, 52
  and inside-out-ness 102–3
Quantum mechanics 81–2

Reading 5, 11, 12, 29, 44, 53, 67–9, 97

Recursion 48
Religion 39, 111, 122, 133, 136
  attitudes to AI 78–92

Repossession (of the past) 3
Resurrection 40–1, 130–2
Retrojection (versus projection) 68–9
Rilke, Rainer Maria 55
Robotics, three laws of 108

Science
  creation 128
  defensiveness of theology 1–3, 5, 83–6
  God of the gaps 22, 91
  and knowledge 82–3
Searle, John 14, 55
Self 3, 6, 31, 41, 45, 53–7, 61–3, 83–7, 111, 114–22, 126–8, 131
  awareness 30–1
  brain activity 36–7
  and consciousness 116–17
  context-specific 120–1
Shakespeare, William 131
Simulation 66, 69, 99
Sin 50
Sliding Definition Ploy (SDP) 26, 57–8, 61–2, 115
Soul
  human 3, 6, 46, 85
  android 115–16
  differing views of 39–41
  generation of 124–30
  hardware/software analogy for 118
  surviving death 119–23
Sperry, Roger 97
Spinoza, Benedict de 43
Spirit, human 6, 85, 92, 116–17, 129
Star Trek 122
Suffering and pain 91–4

Story
    inside 45, 108
    outside 45, 104
Supernatural explanation 87–8
Superstition 2, 88–9
Supervenience 97

Teilhard de Chardin, Pierre 42–3
Theology 2, 5, 74
    defensiveness about science 83–6
    and power of science 89–91
    central pillar of 103
Thinking 10, 19–21, 35–6, 47, 103
Thoughts 30–1, 34
Thousand cuts, death of 83–6
Tipler, Frank 57, 62

Tolerance 110
Torrance, Thomas F. 84
Tradition, Judaeo-Christian 128
Turing, Alan 13–15
Turing machine 12–14, 97, 99
Turing test 14, 26, 30–1, 104, 134

Vagueness
    of performance 77
    of words 23, 57
Verne, Jules 7
von Neumann, John 17

Weizenbaum, Joseph 21, 120
Wittgenstein, Ludwig 36
Writing 35, 52–5, 67
    see also Reading